RADIO CONTROL SYSTEMS

BY FRED M. MARKS

On the cover: Jennifer Ritchie, Marie Kiser, and James Marks plan another flight. Photo by Frank Pierce.

Editor: Burr Angle Art Director: Lawrence Luser

Editorial Assistant: Marcia Stern Artist: Bill Scholz

KALMBACH BOOKS

Published by Kalmbach Publishing Co., 1027 North Seventh Street, Milwaukee, WI 53233. Printed in U.S.A. Library of Congress Catalog Card Number: 80-81430. ISBN: 0-89024-550-9.

1 Basics

In 1898, Nikola Tesla, better known as the inventor of the alternating-current electric motor, successfully demonstrated a radio control system in a model boat. In the 1910's and 1920's the U.S. Army and Navy experimented with radio control for full-size and miniature aircraft. In the 1930's civilian modelbuilders such as Walter and William Good, Ed Lorentz, and Howard McEntee developed RC aircraft that could be built and flown by hobbyists. RC flying events became part of the annual national competitions sponsored by the Academy of Model Aeronautics.

In the 1940's and 1950's the hobby grew slowly, handicapped by a shortage of suitable components and the FCC's insistence that RC modelers be licensed amateur radio operators. The FCC later authorized model operations on several bands of the examination-free Citizens Radio Service, but the hobby still grew slowly because equipment was expensive, bulky, and unreliable.

In the 1960's transistors and printed circuits replaced vacuum tubes and hand-wired circuits. Receiver designs changed from superregenerative to superheterodyne, offering better selectivity, and transmitters shrank until they no longer had to be powered by automobile batteries. In the 1970's integrated circuits began to replace discrete components, bringing a thousand-fold improvement in reliability. As systems improved, they also became less expensive: A system today costs less than its less-capable equivalent of 10 years ago.

Legal requirements

Except for toy radio control units that have an output of less than 100 milliwatts and that operate on specially assigned frequencies, you must have a license issued by the Federal Communications Commission to operate RC systems in the United States. You must be a U.S. citizen, at least 12 years old, and you must have read Part 95 of the FCC rules governing RC operations. No examination is required, and, at this time, no fee is charged. If you buy a new system, the FCC license application should be packed with the set. If you buy a used system, you can obtain a license application by writing the FCC, Washington, D.C. 20544. Obtain the Part 95 rules from the Government Printing Office, 710 North Capitol Street, Washington, D.C. 20401. The cost is $2.00.

Licensing requirements in Canada are similar to those in the United States. Obtain applications from any office of the Department of Communications. In other countries, RC licenses are issued by the department or ministry that regulates radio uses.

Table 1-1 lists the frequencies now used by RC modelers in the U.S. and Canada. Note that the 53-MHz frequencies are for amateur radio operators only, and that most of the 72-MHz frequencies are for model aircraft only. The AMA is petitioning the FCC for additional frequencies. If any changes are made, you'll read about them in the modeling magazines.

The colors listed beside each frequency are those used on flags attached to the transmitter antenna. Always use the correct color for your frequency and always check to see if another modeler is operating on your frequency before turning on the transmitter: Only one modeler can use a frequency at a time.

Do not operate RC models without adequate insurance. You can purchase a rider to your homeowner's policy or you can join the Academy of Model Aeronautics (in Canada, the Model Aeronautics Association of Canada). Membership in the AMA includes a $1,000,000 insurance policy, considered the minimum should a model injure someone.

Because this book is about RC systems, not about building and operating RC models, an extended discussion of safe operating practices is not ger-

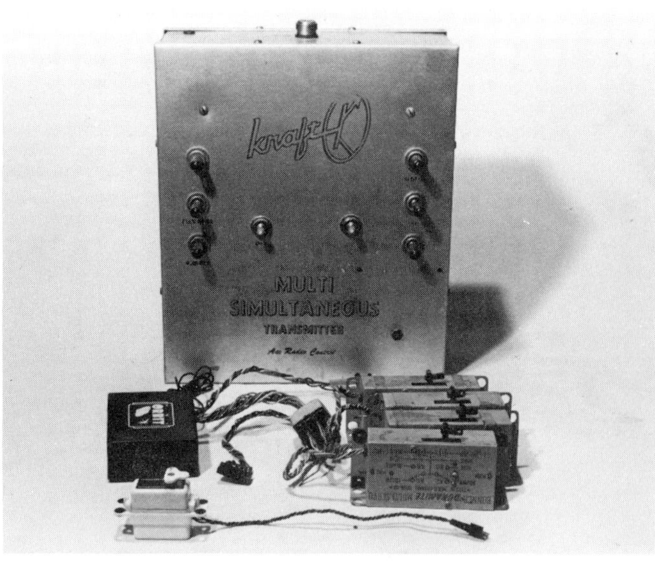

This early Kraft transmitter required 135 volts and weighed about 6 pounds. The Bonner Duramite servos used tuned reeds and were nonproportional. They had only three output positions: all the way out, all the way in, and centered. Compare their size to that of the Ace R/C Micro servo at bottom left.

The Dee Bee Quadruplex used pulsed tone AM to achieve five-channel analog proportional control.

MHz	Color		
26.995	Brown	27.145	Yellow
27.045	Red	27.195	Green
27.095	Orange		
53.1#	Black and brown		
53.2#	Black and red		
53.3#	Black and orange		
53.4#	Black and yellow		
53.5#	Black and green		
72.080*	White and brown		
72.160	White and blue		
72.240*	White and red		
72.320	White and violet		
72.400*	White and orange		
72.720*c			
72.760*c			
72.800*c			
72.840*c			
72.960	White and yellow		
75.640*	White and green		

Amateur radio operators only
* Aircraft only
c Canada only

Table 1-1 RC FREQUENCIES AND FLAG COLORS

mane. Nevertheless, let me state that you must always operate RC models prudently for the good of the hobby and for your own protection.

Choosing an RC system

Some years ago, Jim McNerney, Duane Lundahl, and I wrote perhaps 60 reviews of RC systems for a modeling magazine. Naturally, people often asked "Which radio is best?" In all honesty, we had to answer that, with few exceptions, all the sets performed acceptably. In the next chapter, you'll learn that all digital RC systems are, functionally, virtually identical, so you may become confused. But remember that all automobiles are functionally the same!

I recognize that your first RC purchase will have a bit of emotionalism to it as you convince your parents, spouse, or yourself that you are making a reasonable investment. I'll help you choose wisely by leading you through the following steps:

- Study the kinds of systems on the market.
- Learn how performance standards are stated.
- Investigate the availability of service.
- Examine the advantages and disadvantages of kit-built systems.

Standard systems

A multi-channel RC system consists of a transmitter equipped with various control sticks, switches, and levers; a receiver/decoder; servos; and associated wiring and batteries. If the system uses rechargeable batteries, a battery charger will be supplied.

Single-channel digital sets are available in Europe, but in North America the following are more or less standard: two channels, three channels, four channels, five channels, and seven channels. Systems with eight or even nine channels are less common. Figure 1-1 illustrates typical control stick arrangements. Transmitter control sticks may move in just one direction, or, by the use of gimbals, may move in two or three axes, controlling the same number of channels. The transmitter may also have a two-position switch for retractable landing gear and single-axis levers for auxiliary functions.

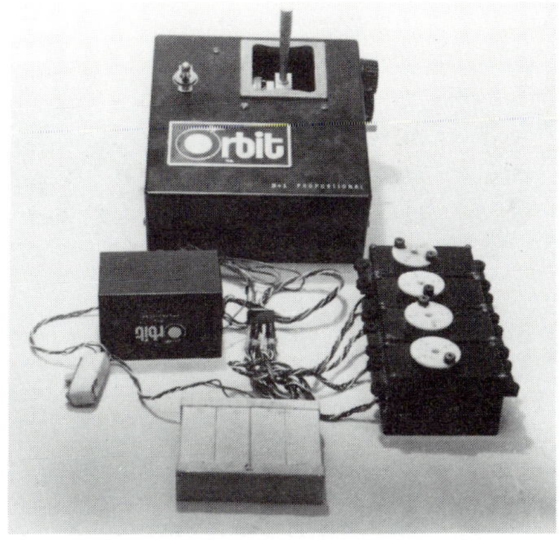

The Orbit four-channel analog proportional system was the next step beyond the Quadruplex. Orbit servos used feedback to help control jitters, but were huge by today's standards.

Systems are available with:

- Two channels with two single-axis sticks, one horizontal for steering, and one vertical for throttle or elevator. Avoid these if you intend to move on to more complex models or if you expect to receive flight training from more advanced modelers who don't use that kind of rig.

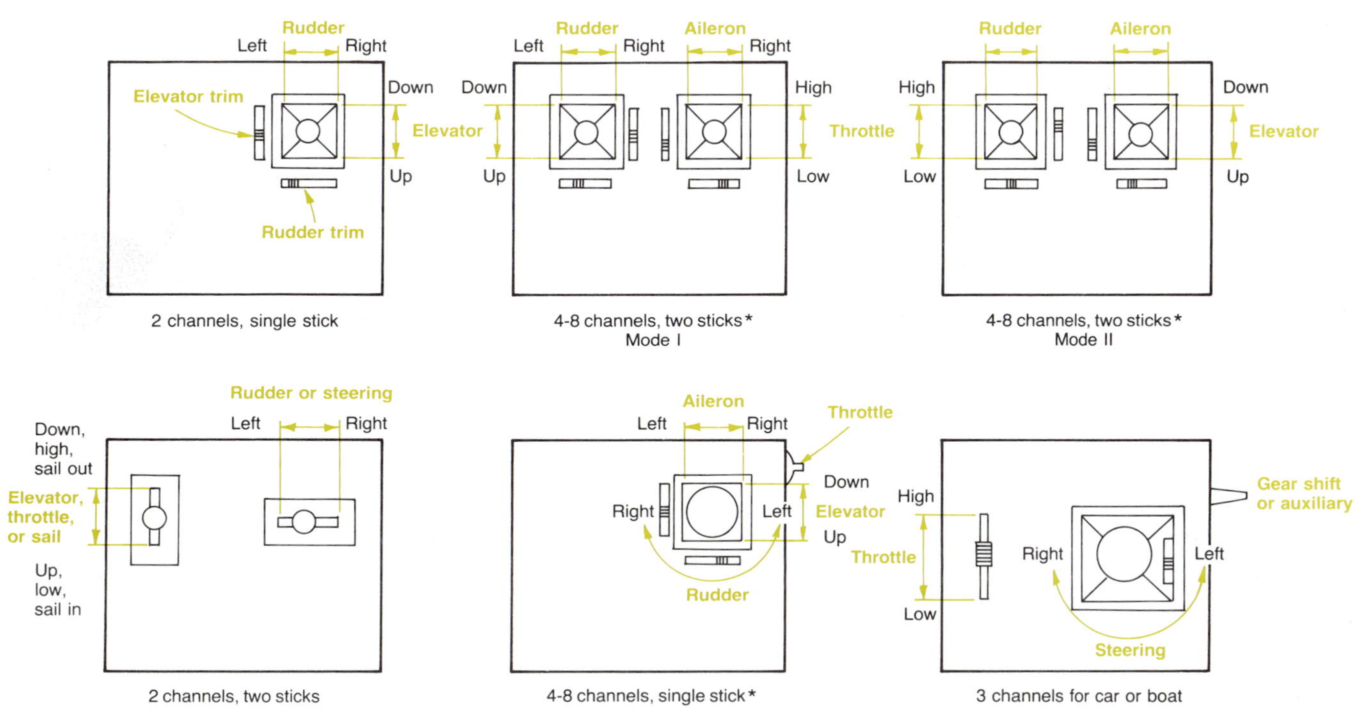

Fig. 1-1 TYPICAL CONTROL STICK ARRANGEMENTS

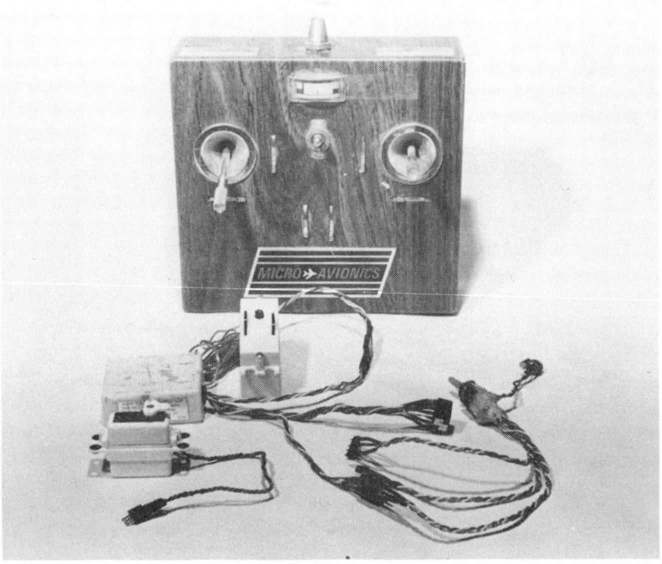

The digital method of modulating control information made this early digital proportional set from about 1964 more reliable than its analog predecessors.

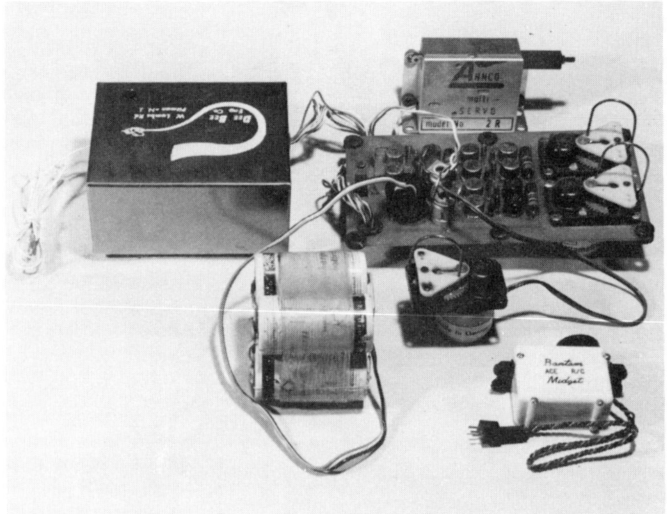

The Quadruplex system of 1960 used the three kinds of servos shown here above the Bantam Midget servo of 1980. The Quadruplex servos had no feedback, so they dithered constantly. Their output torque was less than one-fourth that of the Bantam Midget at lower right.

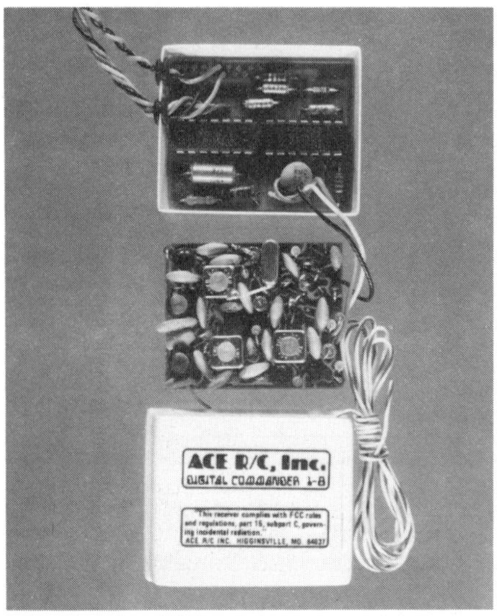

This modern eight-channel receiver uses integrated circuits in the decoder.

● Two channels with a single two-axis stick or a special arrangement with a wheel for steering and a lever for throttle and brake. The single two-axis stick is for gliders and simple airplanes, the wheel-plus-lever style is for cars and boats.

● Three channels with a two-axis stick and a lever. The lever is for throttle on powered models and for spoilers on gliders.

● Four channels with two two-axis sticks for general sport use. These systems can be used for gliders, sport airplanes, special ground vehicles, and, although they are somewhat awkward, for cars, boats, and other models. The usual configuration, for airplanes, has three self-centering controls for ailerons, elevator, and rudder and one positionable control for throttle.

● Five channels with the same arrangement as the four-channel but with a switch for the fifth channel to control retractable landing gear. This is the most popular configuration and has the best resale value.

● Seven channels with the same arrangement as the five-channel set plus two levers for auxiliary controls. Pilots of competitive scale aerobatic aircraft and skippers of fancy boats can often use seven channels.

These systems have a great deal in common since the advent of integrated circuits, especially the Signetics NE5044 encoder, the NE5045 seven-channel decoder, and the NE5046 two-channel decoder. The NE5044 and NE5045 can be used equally well in three- to seven-channel systems. Also, a manufacturer often uses the same receiver and decoder for all systems with more than two channels. (Most two-channel systems still use discrete components, so are a special case.) He

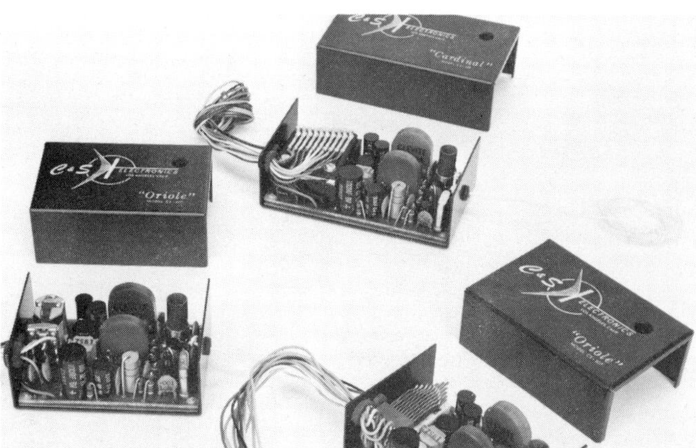

Three superheterodyne receivers with crystal filters from the mid-1960's. The unit at left is a single-channel receiver with a relay. The other two units used tuned reeds that responded to audio tones of 300 to 600 Hz to drive transistorized servo amplifiers.

The gadget to the left of the servo is a compound escapement from the 1950's. It was powered by a twisted rubber band attached to the hook. One push on a switch on the transmitter kicked the rudder all the way right, two pushes sent it all the way left. Rubber band was wound up for each flight.

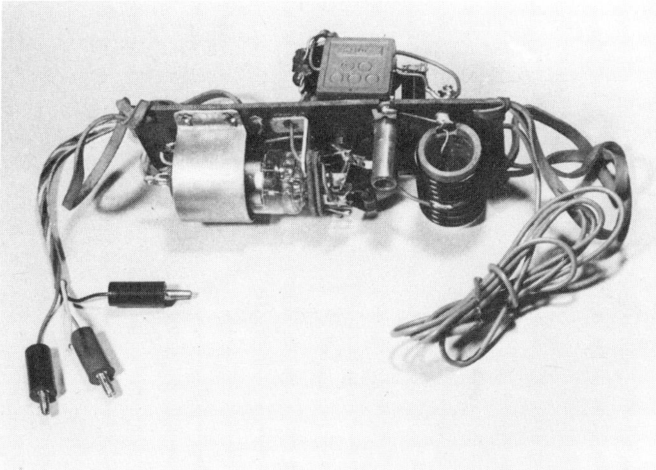

Here's a real old-time receiver. The unit was suspended on rubber bands to protect the vacuum tube and relay from engine vibration. Even so, it had to be tuned after each flight.

This receiver used a big relay inside the case to actuate a rubber band-powered escapement that moved a control surface.

supplies the same servos with all systems. The same transmitter electronics are frequently used for all systems from three to seven channels, with any unused channels left blank. Batteries, chargers, wiring harnesses, and accessories are usually the same no matter how many channels are used.

Super systems

Standard systems will fill the bill for most modelers, but if you demand more flexibility, consider one of the "super systems." Super systems such as the Kraft Signature Series, the Royal Omega, and the Ace Silver Seven contain some or all of these special features: buddy box connections, servo reversal, adjustable servo throw, dual-rate selection of servo motion, nonlinear servo response, programmable functions, electronic mixing, and sophisticated metering. Note that the special features are in the transmitter: The airborne unit, almost without exception, is a standard unit.

The buddy box feature will appeal to you if you are a beginner, if there are experienced modelers to train you, and if your instructor's system and yours are of the same make and series. The buddy box functions as follows: a) The two transmitters are the student's and the trainer's. b) A 6′ cable with a connector at each end links the two transmitters. c) The trainer's end of the cable has a shorting connection that kills the RF output from the student's transmitter, while permitting the trainer's transmitter to radiate. d) The cable permits the trainer to control the model via the encoder in his transmitter until he depresses a spring-loaded deadman's switch to enable the student to control the model through the student's encoder. e) The trainer may re-

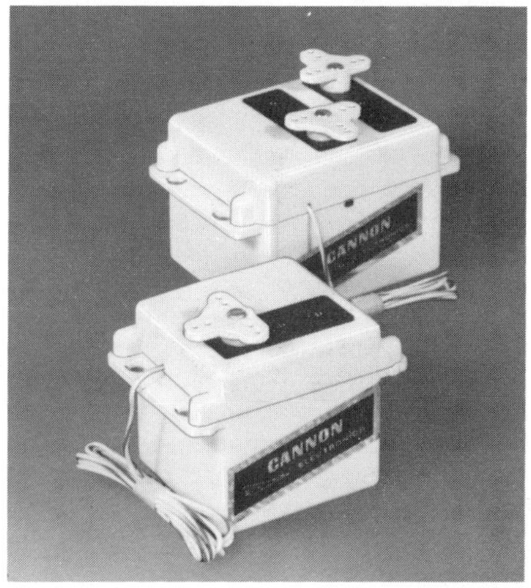

If the receiver contains one or more servos in the same case, it's called a "brick." Bricks are often used in gliders and aircraft with narrow fuselages.

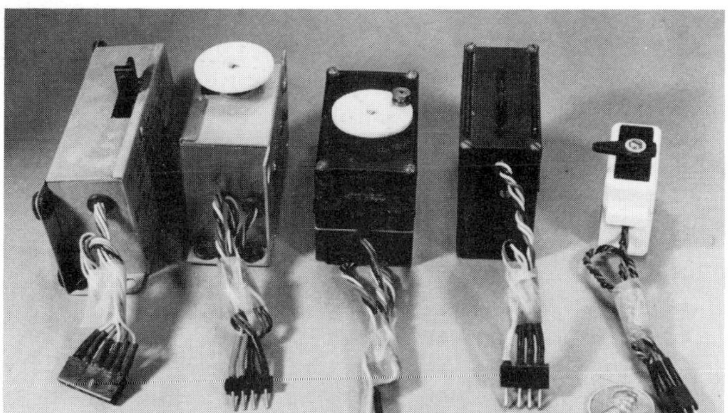

From left to right, a Bonner Duramite reed servo, a Sampey analog proportional servo, an Orbit "boxcar" servo, an early World Engines digital servo, and a tiny contemporary Ace Micro servo. Except for the Duramite, each of these servos has about the same output torque.

In a single-deck receiver (left) all components are mounted on one PC board. A two-deck receiver (right) uses a separate PC board for the decoder. Both are the same electronically.

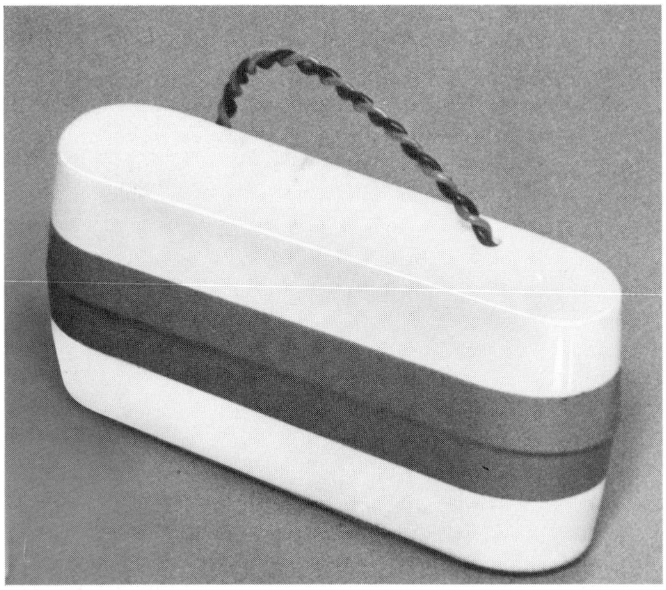

These nickel-cadmium battery packs for the airborne portion of the RC system differ only in shape, so you can use whichever fits best in your model.

sume control instantly by releasing the deadman's switch. Having taught both my sons to fly, one without and one with the buddy box, I can attest to its usefulness. Progress is faster and tension is less when the buddy box is used.

Servo reversing in the transmitter is especially useful when setting up the system in a new or different model. It is most effective when used in conjunction with variable throw adjustment for the channel of interest. For example, the Ace Silver Seven permits servo reversing on four channels and throw adjustment over a broad range on all channels. My first experience with these features came when I installed the system in a Goldberg Falcon 56 MK II, a four-channel airplane. Previously, with neither of these features, it would have taken an hour or two to move pushrods, reverse controls, adjust linkages, and set up the model for flight. The entire sequence of selecting servo throw and travel with the new transmitter took less than 5 minutes.

The servo travel adjust is particularly useful for the throttle servo. We used to have to use throttle override devices and fiddle with the linkage to keep the throttle settings within the stops so that the servo wasn't overloaded at either end of travel. With travel adjust, the throttle end travel can be precisely set at the transmitter. Keep in mind that if the same transmitter is used for more than one model, you must reset the throw direction and range of travel when switching models.

Dual-rate selection, or more appropriately, dual servo travel range selection, is a popular feature for precision aerobatic fliers. Dual-rate is normally featured on the first three functions (ailerons, elevator, and rudder) and can be selected by a switch for each function. A precision flier can switch to low range for the exact amount of elevator travel needed for high-speed maneuvers such as loops, but can switch back to high range for full control authority during takeoffs and landings, and for low-speed maneuvers such as spins. Both high and low range are adjustable, with low range for some units accessible during flight.

Dual-rate is especially convenient for pilot training. The instructor can have full control authority for takeoff and landing, but can switch to low range when the student takes over to prevent the usual student tendency to overcontrol. Not as apparent is the application to gliders, where the flier has to be cautious of the galloping behavior of gliders in level flight. Use high range during tow and maneuvers and switch to low range as the glider enters a thermal. Use high again for landing.

Nonlinear control is the primary alternative to dual-rate. As the stick first moves away from neutral the slope is relatively low, but it rises smoothly to a higher slope after a preset stick deflection. This means that the flier or driver can have smooth, tight control during the straight portions of a flight or run, and yet automatically have full control authority for turns. Nonlinear control requires some acclimation if you've flown for years with linear control, but once you're used to it, it's a real boon.

Programmable functions can be used for precision aerobatics or just for horsing around. A pushbutton applies a precise, predetermined amount of control as long as the button is held down. That control setting is adjustable during flight to any desired position. Programmable controls are often used to help perform aerobatic maneuvers such as slow rolls when the flier must coordinate rudder, elevator, and throttle to achieve a competitive maneuver. That's like scratching, chewing gum, and rubbing your head while kissing your lady! The flier's job is easier if he can punch in the desired roll rate and concentrate on the other controls. Of course, programmable functions can be set up for loops, snap rolls, spins, and whatever else your imagination allows.

Electronic mixing of two control movements is useful for aircraft with V-tails, flaperons, or coupled flaps and elevator. The mixer is particularly convenient for helicopter pilots, who need three hands to fly. The idea is to have tail rotor pitch change automatically by mixing the throttle or collective pitch with the tail rotor pitch control. As throttle or pitch is increased, tail rotor pitch increases to prevent yaw (or pinwheeling, when the beginner loses it!). Additionally, unidirectional mixing of throttle position into collective pitch makes flying the helicopter easier. My experience with electronic mixing goes back several years to coupled flaps and elevator in a Topflite RC Nobler and earlier experiments in an Andrews Trainermaster. After several attempts to use mechanical mixing in these models, I can report that electronic mixing is superior. It's difficult to make a lightweight but reliable mechanical mixer.

Sophisticated metering can take the form of an expanded-scale voltmeter that accurately measures the battery voltage in the transmitter. Some units also have an RF meter. Some let you plug the expanded-scale voltmeter into the airborne battery pack to measure its voltage under load. Others use a low-voltage detector to signal via a light-emitting diode when safe voltage

This Ace Micro servo with an IC servo amplifier is hardly larger than a dime, but it produces 4 pounds of thrust and rotates through 90 degrees in just half a second.

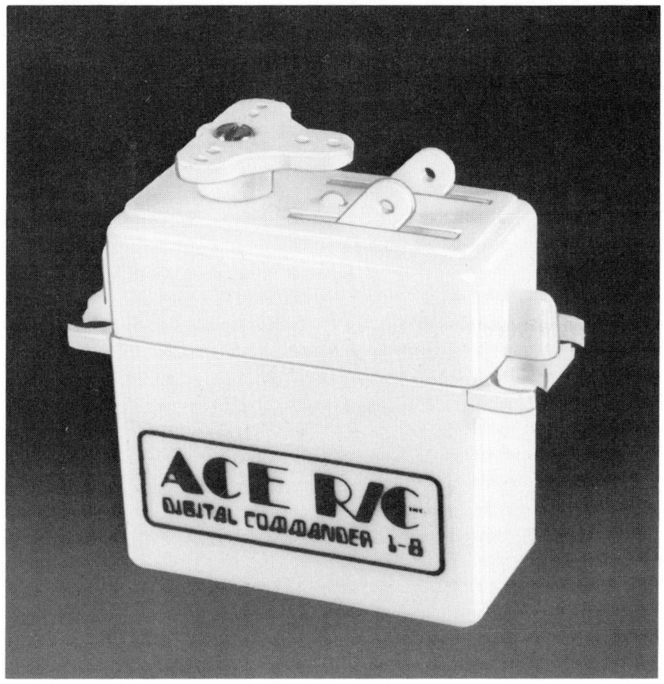

A multiple-output servo lets you choose between rotary and push-pull movement.

bounds are exceeded. One set even includes a flight timer with a buzzer to warn you that the fuel tank is almost empty.

Some modelers have asked for dual-sensitivity control trims so that a new model could be operated with a broad range of trim for initial flights. That feature has been included in the Ace Silver Seven transmitter. Once linkages are adjusted to give the proper neutral setting without trim, the broad trim is no longer needed, so the pilot switches to low trim throw.

One or two sets feature an arrangement that permits a cable from the buddy box plug to be connected to the decoder so that the servos can be exercised without turning on the transmitter. That's useful at a contest or other busy site when adjustments are needed and you can't tie up a frequency without getting your model stomped.

Most transmitters have metal and injection-molded plastic control sticks. However, precision-machined all-metal stick assemblies are more rigid and allow the use of larger control potentiometers for more linear response. These do require separate electronic trims and add nearly $100 to the cost of the transmitter.

Other features include an antenna that collapses completely into the transmitter. I don't trust these to retain a solid connection over years of use. Some transmitters have an internal battery charger that charges both the transmitter and receiver battery packs, usually both at the same time. I prefer a separate transformer-isolated dual-output charger that can charge the transmitter and receiver independently. Such a charger can also be used to charge spare battery packs without tying up the transmitter.

Take heed: No matter how many features are listed above, there's always someone who wants "just one more." That's okay, because that's what innovation is about. Still, I refuse to consider the proposal for a microprocessor that would produce a precision aerobatic pattern; it might even work — when there's no wind.

System specifications

You should know how system specifications are stated and what they mean. The measures of performance that follow are the most important.

Transmitter RF output is specified as "RF into the final," which means that an RF voltmeter is placed across a precision resistor in the input connection to the final RF amplifier transistor. This figure generally ranges from 600 milliwatts to 1000 milliwatts (1 watt) in the 72-MHz band. A more appropriate measure of power output is radiated power as measured in the field (over 100 feet away) on a calibrated range as required for FCC type acceptance. This figure ranges from 250 to 600 milliwatts in the 72-MHz band. If input to the final is specified, the radiated power can be estimated to be 55 percent of that. New designs using depleted metal-oxide silicon and vertical metal-oxide semiconductors offer efficiencies approaching 70 percent.

Transmitter frequency tolerance is rigidly specified by the FCC at .005 percent in the 72-MHz band. You may assume it is achieved. Some manufacturers promise a tighter tolerance, say .001 percent, for systems on the 27-MHz band with interchangeable crystals.

Encoder frame rate and control pulse width are important if you plan to use a second airborne pack or interchange servos and receivers. Frame rate averages about 20 ms. Even more important is control pulse width. Most sets use 1.5 ms $\pm$ 0.5 ms for full control. Servos and receivers should have approximately the same specifications if they are to be interchanged.

Transmitter battery voltage depends upon the type and number of cells used. Most transmitter battery packs are 9.6 volts, eight nickel-cadmium cells hooked up in series. A few are 6 volts, five nickel-cadmium or four alkaline cells connected in series. Many low-cost sets use a 9-volt carbon-zinc battery. Others use six alkaline cells for 9 volts, while a few use eight such cells for 12 volts. The voltage used is important if you plan to use a battery cycler, which must be matched to the battery voltage.

Nickel-cadmium battery capacity is stated in milliampere-hours (mAh). Most transmitter nickel-cadmium battery packs are rated at 450 mAh or 500 mAh. Some have 550-mAh cells (derated sub C cells). Transmitter weight and balance are adversely affected by heavier cells, although operating time may be increased.

Size and weight are also important for the airborne battery pack. The smallest nicad packs are now about 2 ounces for a four-channel/four-servo system and have a capacity of 225 mAh. Mid-size packs are 4 to 6 ounces with 450-mAh cells. Large packs for use with heavy-duty servos are rated at 550 mAh and weigh 12 to 16 ounces.

Although rechargeable nickel-cadmium batteries were standard for a long time in both the transmitter and receiver battery packs, the use of alkaline cells has recently become widespread in low-cost two- and three-channel systems. The alkaline cells

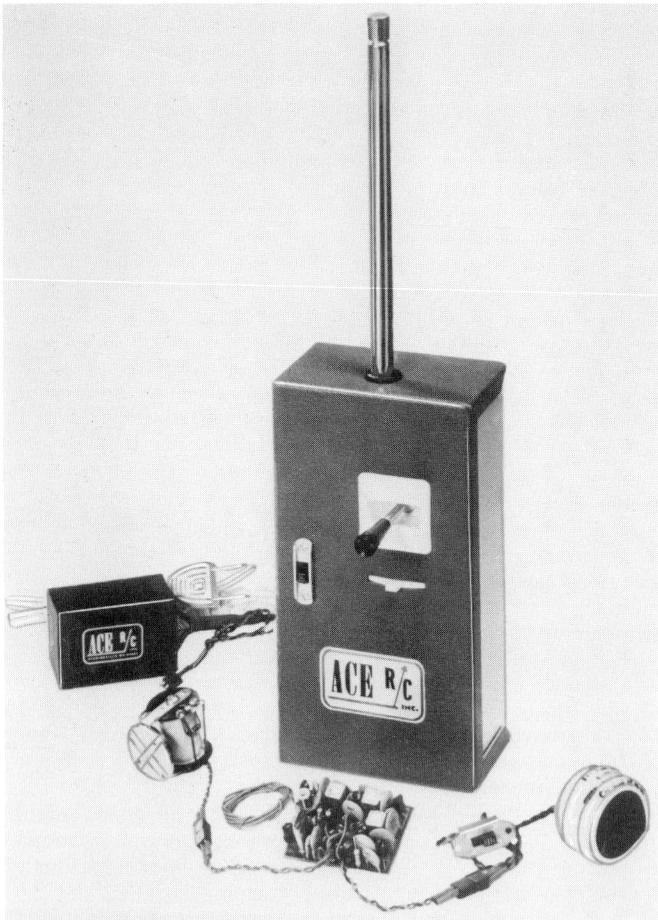

This Ace single-channel pulse proportional system has survived the digital revolution because it is lightweight, reliable, and inexpensive.

This Cannon Electronics two-channel system has a brick receiver with one servo, a separate servo, and a 225-mAh nickel-cadmium battery pack for the airborne unit. It's the perfect choice for a glider or boat.

cost $5 to $6 a set and must be replaced at least once a season, more often if the system is used a lot. Be aware that nickel-cadmium battery packs and chargers cost as little as $25 and last for years. It's usually cheaper to start with nicads and a charger.

Receiver sensitivity is specified in "microvolts for full decoder control," and ranges from 1.5 to 3.0 microvolts for most receivers. If it's less the receiver is too sensitive; if it's more the manufacturer won't admit it. How important is this figure? Under normal conditions, it's not as important as is frequently stated. A manufacturer once told me he often flew safely with a receiver that had only 10-microvolt sensitivity. This is true — under normal conditions — but I wouldn't want to fly a model whose receiver had less than 3- to 5-microvolt sensitivity unless I also had a very powerful transmitter. For example, the combination of a transmitter with only 350-milliwatts output and a receiver with only 5-microvolt sensitivity would not be good.

Receiver image rejection, which is determined by front end design, is specified as "X decibels down." This means that the receiver front end will reject signals on the image frequency that are X decibels in amplitude above the signal level at the desired frequency. This figure isn't always specified because it's sometimes close to zero. That means the front end will respond equally to the image frequency and the desired frequency, which is bad. A good figure for image rejection is 9 to 16 decibels or higher.

Receiver selectivity is defined in terms of "3 dB down at $\pm$ X kHz," where X is usually 2.5 to 7 kHz from the center RF. Receivers with a selectivity of 3 dB at $\pm$ 2.5 kHz reject all signals 3 dB lower than the desired frequency as long as they are at least 2.5 kHz above or below the desired RF. Of course, the receiver cannot reject signals that are within 2.5 kHz of center, which is why RC operation in the 27-MHz band was devastated when some CB'ers began to use amplifiers with several hundred watts output. Even the skirts of such illegal transmissions are more powerful than the RF received from our RC transmitters. Complete jamming results; neither front end nor IF, or even double conversion, gets around the jamming. Receivers with IF transformers have 3-dB bandwidths of $\pm$ 2.5 to $\pm$ 3 kHz, while those with crystal filters have at least $\pm$ 3.5-kHz bandwidth. However, the skirts of the crystal filter are sharper, so that they may be 16 dB down at $\pm$ 4 or 5 kHz, while the transformer IF is only 6 or 8 dB down at that bandwidth.

Automatic gain control dynamic range is important, particularly in aircraft where distances and altitudes change rapidly. As signal strength fluctuates, the AGC must be able to follow quickly enough to maintain constant overall receiver gain. Since the AGC is a ladder filter, its time constant is usually a compromise. AGC dynamic range is specified in terms of "linearity from full limiting to no AGC." It is difficult to be specific about the amount of dynamic range needed. A range of 90 dB or greater is outstanding, while less than 60 dB is cause for suspicion.

Servo resolution is defined in terms of percent error or as microsecond error response. Percent error is a reasonable way to state servo performance, but it is more meaningful to state performance in fractional degrees of full servo rotation. Servo resolution error of more than $\pm$ 0.5 degree is poor, but not unacceptable for the sport modeler. Precision aerobatic pilots demand the best servo resolution they can find.

Servo output force should be stated in inch-ounces of torque. This means that the servo produces *X* pounds of thrust at a given radius, as from the center of the output arm or wheel to the control link connection. Most output wheels have inner holes at about 0.25″ radius and outer holes at about 0.4″ radius, but there is considerable variation. Small servos such as the Ace Bantam Midget or Micro Bantam produce 20 inch-ounces. Larger servos such as the Bantam high-thrust and Kraft KPS-14 produce about 30 inch-ounces. Some of the super servos such as the Royal Chevron and Ace Ultra produce 40 to 50 inch-ounces. Your choice of servos is determined primarily by the size of the model and the anticipated control surface forces. If linkages are smooth and free of binds, and if control surfaces are properly balanced statically and dynamically, then small servos can control even large models, up to .60-powered aerobatic planes. If the linkages bind and control surfaces aren't balanced, even the largest servo won't operate properly.

Servo transit time is defined in fractional seconds for end-to-end movement of the output wheel. Travel time of 0.5 second is good, 0.4 is better. Transit time greater than 0.6 second is too slow.

End-to-end system linearity and repeatability are true measures of overall performance. Now that nonlinear control is possible, repeatability is even more important than it was. Lack of repeatability and unplanned nonlinearity are introduced by control stick flexure and poor servo resolution. It does little good to specify linearity without also specifying repeatability, though improvement of either adds to system performance.

Service

Sooner or later the system will need to be serviced. This inevitable requirement may be met in four ways:

- The set may be repaired by the manufacturer's service center.
- The set may be repaired by a certified regional service center.
- The set may be repaired by a qualified, but not necessarily certified, local serviceman.
- The set may be repaired by its owner.

For now, I'll limit discussion to the first three sources of service. All manufacturers have a well-equipped service department and they usually have competent technicians. Recognize, however, that the manufacturer must compete for technicians with aerospace and electronics businesses, and that trained, dedicated technicians are always hard to come by. Therefore, there is no absolute guarantee that a factory service center will do good work.

A certified regional service center will have considerable test equipment and will usually have one or two capable technicians. The regional service center may be nearby, so you can ask questions and even "lean on" the technician if necessary.

A local serviceman is a great asset. My heart sinks whenever one leaves my hometown of Washington, D.C., because I can expect numerous calls to service local sets; a task that I've little time to do. The quality of service from a local repairman varies widely. He may have a well-equipped shop and know exactly what he's doing. On the other hand, he may have little equipment and less training. Ask other modelers who have used his services before committing your system to his hands. The local repairman labors under a number of handicaps. He may not always be able to obtain schematics. His test equipment, though perhaps fairly extensive, will be less than a service center's (a good signal generator costs $1500 to $2000). He can't stock parts for more than one line and he doesn't get a good discount on them. Thus, if you plan to rely on a local serviceman, you'd better buy the line he services.

Sets manufactured abroad must generally be sent to the manufacturer for service; rarely is service available from a certified service center or local serviceman, partly because transistors and other components seldom cross-index to U.S. component designations, which makes replacement nearly impossible. Yes, many imported sets work well and some are less expensive than domestically made systems — just keep in mind that service may be a problem.

If a set has been in a crash or malfunctions, the following steps are in order:

This simple, inexpensive Ace three-channel transmitter is fine for sport aircraft. The two-axis stick on the right controls rudder and elevator, while the lever on the left is for throttle.

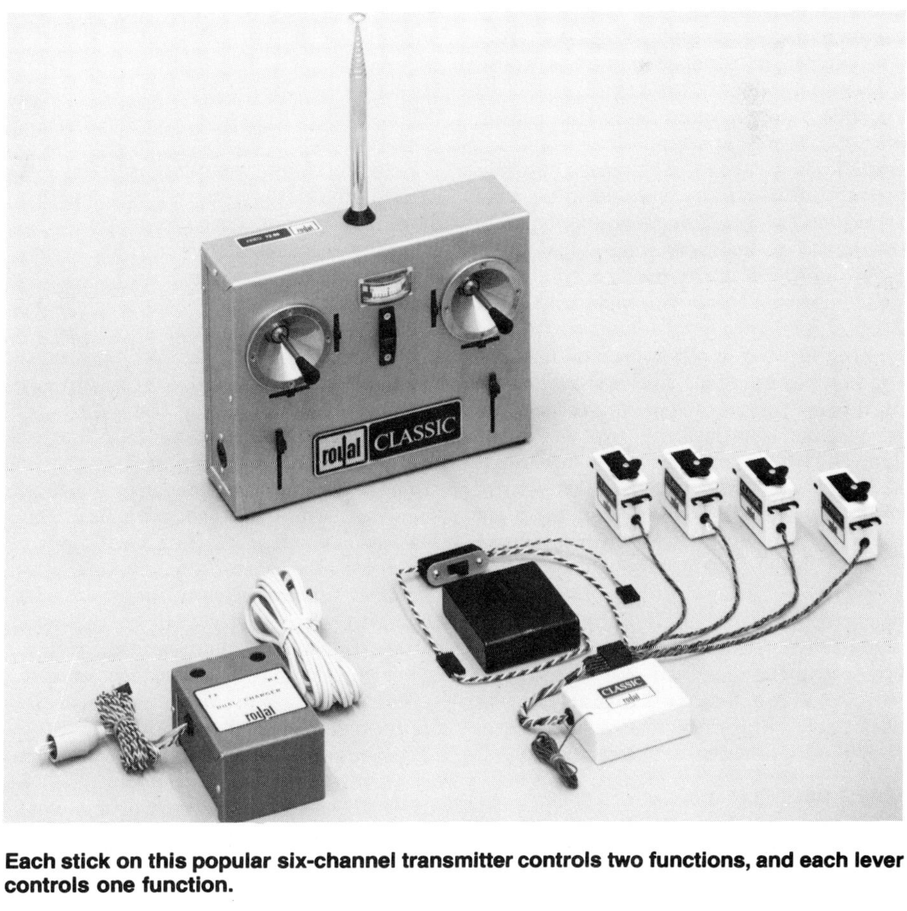

Each stick on this popular six-channel transmitter controls two functions, and each lever controls one function.

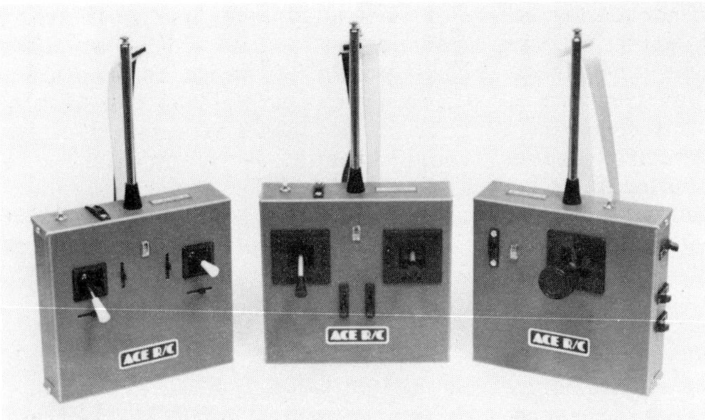

Levers and switches for auxiliary controls may be variously located on the front, sides, or top of the transmitter case. The "single-stick" unit on the right uses one three-axis stick for rudder, elevator, and ailerons.

Replacement servos for this six-channel Hobby Lobby system built by EK-Logictrol are inexpensive because the servo amplifiers are built right into the receiver.

Step 1. Clean the set to remove dirt and oil. Charge the battery packs.

Step 2. Write a letter in which you describe as clearly as possible what seems to be wrong.

Step 3. Give clear instructions about what you want done. If you only want the major problems corrected, state that. If you want a general overhaul in addition to the repairs, state that. If in doubt as to the cost of an overhaul, ask first.

Step 4. Pack the system carefully. If the system came in a foam plastic package, use it. If not, use a sturdy cardboard box at least 3" larger than the maximum dimension of the system. Fill loose spaces with foam plastic or shredded newspaper. Seal the carton with filament tape.

Step 5. Be aware of the service center's billing policy. If repairs are to be paid in advance, include a check that is marked "Not good for more than *X* dollars," where *X* is the maximum charge you've agreed upon or expect to pay. If it's not enough, you'll hear from the technician. If the bill is less, he'll fill in the correct amount. Too many servicemen have failed to receive payment to be caught short again.

Step 6. Unless the problem is quite specific (my dog once chewed up a servo), send in the entire system. The problem may not be where you think it is. I know of a set that was sent to California three times before the owner brought it to me to find the problem, not in the transmitter, as suspected, but in the charger that had never been sent in. Request a full explanation of the repairs and enter these in the system's log when the set is returned.

Used sets

Used sets are potential bargains — and sources of hard feelings, if you aren't careful. There are several guidelines to follow. Is the manufacturer still in business and likely to remain so? Are parts and service still available for that particular system? Is the equipment outmoded, with large servos, a large receiver, and bulky connectors? Wiring gets stiff over the years, is it still flexible? Are any parts broken? Do the nickel-cadmium battery packs need to be replaced? Has the set been serviced regularly? ("Of course," the seller states with an angelic look, "I sent it in every year.") Has the system been modified? (Servicemen double their fee for modified systems.) Has the set been used in racing cars or boats? If so, the servo gears will have taken a beating. Has the set been through a crash? ("Never been crashed," he replies tersely.)

Sets that have seen heavy use by an active modeler are usually not a good buy after about three seasons unless you plan a complete overhaul. Check to see if the manufacturer will overhaul the system and ask for a quote. If you don't hear from him, don't buy the set. If the price is right and the overhaul makes economic sense, you might still be interested in buying the set.

Now ask the seller to demonstrate the system through several flights or runs. If it's in a plane, be sure it's flown to several hundred feet away and to a reasonable altitude. If it's in a car or boat, be sure runs are made at full throttle. Watch for roughness, wandering, and glitches. If you are allowed, have a friend who is a capable pilot or operator try the set or try it yourself.

Naturally, be aware of the system's frequency. Don't buy a set on 50-54 MHz unless you are a ham radio operator. Avoid sets on 27 MHz unless you are in some unknown part of the world where there are no CB'ers. If the set is on 72 MHz, keep in mind that it is illegal to operate ground vehicles or boats on aircraft model frequencies. The frequency can usually be changed by the manufacturer or local service center for a small fee — but check first.

If you stick by your guns and go through the above process, a used set may be a good buy. My first digital set, a used Bonner Digimite, proved to be a fine system with which I learned to fly reasonably well. It still worked well when I sold it to buy a newer set with a smaller airborne unit. I've sold perhaps 12 of my used sets and always give a warranty period, but I don't think the average seller will do that.

Kits

Kits for complete systems are available from Ace, Cannon, Heathkit, and Royal. Why build a kit?

- If meticulous, the kit-builder can produce a superior system — because he built it himself, he has no doubts about the quality of workmanship.
- Schematics, step-by-step instructions, pictorials, and troubleshooting data provided with the kit can make the kit-builder his own future serviceman.
- The skilled builder can customize the set to the degree his talents permit. Be prepared to do your own service if you modify a set.
- The system can be upgraded at any time: A three-channel can be converted to four-channel, and so on.

It takes a fair amount of time to build a set correctly. I estimate that a transmitter takes 8 to 10 hours, a receiver 4 to 5 hours, a servo 1 to 2 hours, and a separate battery charger 1 hour. Assembly time varies with the builder's experience, so treat these times as relative. You'll need a modest assortment of hand tools and a low-wattage soldering iron. Some kits require an oscilloscope for tuning, so if you don't have a scope, don't have access to one, and don't plan to buy one, look elsewhere. As always, check for availability of service before investing in a kit system. Several kit manufacturers offer local service, others do not.

Do not attempt to build a system from a kit unless you know how to solder or are prepared to learn. If you don't enjoy working with small parts, and don't like to follow step-by-step directions, stay away from kits. Finally, you should know that because the quality of workmanship varies from individual to individual, kit-built systems often have lower resale prices than manufactured systems.

Reviews

Magazine reviews can be useful guides when selecting a system. Recognize, however, that such reviews are a trial to a publisher's conscience because the manufacturers whose products he reviews are also his advertisers. The publisher's usual policy is simply to avoid reviewing sets that don't perform satisfactorily. Sometimes, the publisher notifies the manufacturer of deficiencies and gives him a chance to make corrections before reviewing the system. There have been no consumer products type reviews of RC systems by independent laboratories, primarily because such tests are very expensive.

Where to buy

Systems may be purchased directly from some manufacturers but are most often bought at a hobby shop or from a mail-order house.

Several mail-order houses sell a brand manufactured for them by someone else. It's often hard to tell just who the manufacturer is, and whether the system is manufactured in the U.S., Japan, or elsewhere. A few mail-order houses change brands from time to time, so parts availability may be a problem. Be sure to get answers to these questions before buying.

Hobby shops may not be able to offer a discount as large as mail-order houses, but a local hobby-shop owner can provide valuable advice to beginners; he's more likely to give this advice cheerfully if you've purchased the system from him.

Whether you buy the system by mail or locally, determine what warranty comes with the set. What is the service policy? What is the promised turnaround time? The location of service centers is also important. If you live on the East Coast and the set has to go to California every time it needs repairs, that will cost time and money.

Other considerations

Don't buy equipment with limited capabilities just because it's cheaper, particularly if you think you are going to become an active RC hobbyist. Many people buy two-channel equipment, then find little market for it when they want to move up to four or more channels. You may want to special order a four- or five-channel system with only two servos to start. This is a good place to mention that system prices are always quoted with two servos for two- and three-channel systems, and four servos for four- to seven-channel systems. If you can afford it, a four- or five-channel system with four servos is the best buy. It can take you from beginning to advanced status and it has good resale value.

The "plain Jane" systems in most manufacturers' lines usually have the same electronics as the rest of the line. The Kraft Sport Series is one exception: These systems use a different frame rate and pulse width than other Kraft products. The EK International Champion (also sold as the Hobby Lobby) was another exception. It is unusual in that all servo amplifiers were in the receiver. The servos were inexpensive but you had to accept the fact that the entire receiver must be sent in for service of a servo amplifier.

Interchangeability

It is practical to mix-and-match equipment from different manufacturers — within limits. The biggest obstacle to matching servos and receivers is incompatible connectors. You can solve this problem by chopping off the old connectors and substituting Deans connectors. Of course, receivers and transmitters must operate on the same frequency and the transmitter and receiver must be tuned to operate together. Servos may have to be recentered. Ace and others sell airborne packs designed to be compatible with all digital transmitters. Similarly, a new transmitter can be matched to an existing receiver. Servos made by Ace, Royal, and Litco are designed to work with most decoders. Servos from other makers may or may not be interchangeable; ask the manufacturer before purchasing.

In summary, choose an RC system on the basis of your needs, the amount you wish to spend, and the availability of service. Size, weight, and special features are secondary, but worth investigating. Most sets on the market are acceptable and the old-line manufacturers are dependable. Ace R/C and World Engines have been in the RC business the longest, followed by Kraft, Heathkit, Cannon, Royal, Pro-Line, and Westport International.

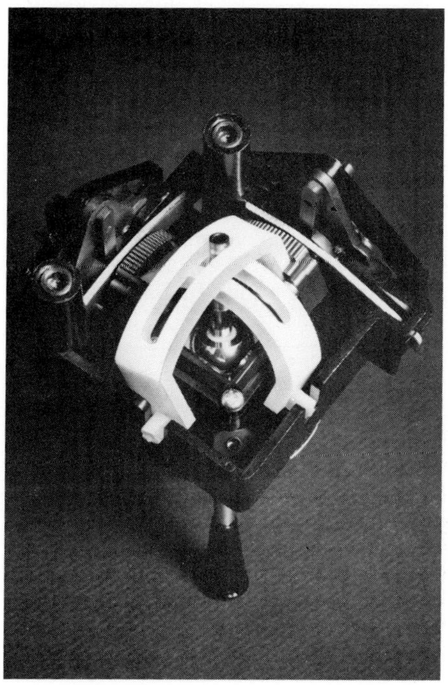

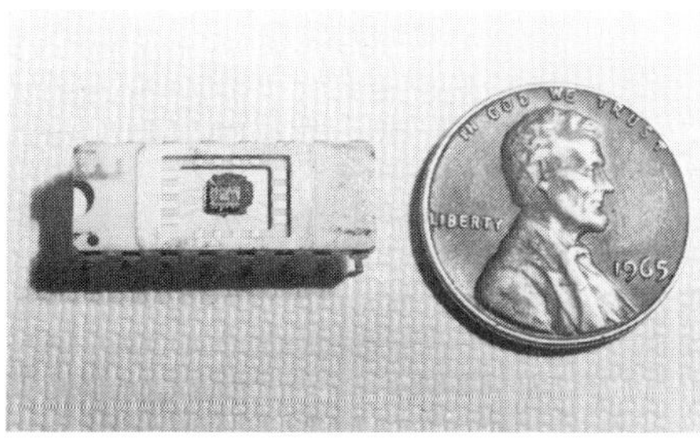

Remove the cover of the NE5044 IC package and here's what you'll see. The active circuitry is the little dark blob in the center of the recessed area.

(Left) The precise control response of this closed-gimbal stick assembly is made possible by high-strength plastics and modern injection-molding.

(Right) Signetics NE5044 integrated circuit in the encoder of the Silver Seven transmitter makes possible such features as nonlinear servo output movement and switchable servo reversing.

2 Transmitter design

The next three chapters tell how RC systems work. I'll describe only the two types of systems currently in use and I'll keep theory to a minimum because that's primarily of interest to system designers. In discussing digital proportional systems, I'll explain general principles of operation, illustrate these by examples of past and present systems, and then describe the most interesting features of a typical super system that uses state-of-the-art integrated circuits. This chapter first provides general information about RC systems and then explains in detail how the transmitter in a digital proportional system encodes and transmits control information.

In any radio control system for models, the objective is to translate the motion of a control stick on a transmitter into motion of a control surface on a model. Radio control systems do that in this way: An encoder translates information about the position of the control stick into electrical pulses. This encoded information modulates a radio frequency carrier wave produced by the transmitter's RF section. The modulated RF wave radiates through space to a receiver/decoder in the model which demodulates and decodes the encoded information. The decoded information passes to electromagnetic actuators or motor-driven servos that move the control surfaces, Fig. 2-1.

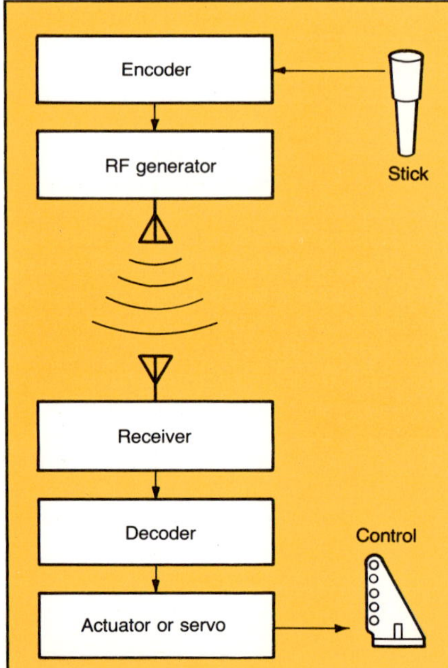

Fig. 2-1 BASIC RC SYSTEM FUNCTION

Pulse proportional system

In the Ace pulse proportional single-channel system the transmitter communicates the encoded information to the receiver which demodulates and decodes the information before passing it to an electromagnetic actuator that produces the driving force on the control surface. Figure 2-2 shows the principle of the actuator: The bar magnet in the actuator always attempts to keep its north pole aligned with the south pole of the electromagnet. But the electromagnet changes polarity each time the finger moves the common wire from +2.4 volts from one contact to the other. Replace the finger and the wires with the transmitter and receiver and we have the situation in Fig. 2-3. It's not easy to push the button on the transmitter, Fig. 2-2, so this chore is performed by the pulser built into the transmitter and controlled by a stick. Figure 2-4 shows how it operates. The pulser keys the audio oscillator in the transmitter 4 to 6 times per second so that the rudder flaps back and forth at the same rate. In flight, the airplane doesn't respond to each wag of its tail, but does respond to the average position of the rudder. If the rudder spends more time on the right side than the left, the airplane turns to the right, and vice versa.

Pulse position modulation

In order to avoid a possible misunderstanding, you should know from the outset that our model RC digital proportional systems are not true digital systems as that phrase is used in electronic engineering. In electronic engineering, digital systems are most often used to transmit analog data (usually a human voice) in which the amplitude of a control voltage (a reflection of the amplitude of the voice) determines the time period between succeeding pulses, Fig. 2-5. Decoding is easy because a resistance-capacitance integrator reconstitutes the original signal. This modulation scheme is called pulse position modulation (PPM).

This type of digital control has not been developed for models because the digitalized input would have to be generated by shaft encoders or other analog-to-digital converters (one per channel) and similar devices would be required for decoding. Encoding and decoding equipment is so large and expensive that PPM is impractical for models.

Model digital systems

To the best of my knowledge Gerry Pullen made the first experiments with the present concept of modulating information for model digital proportional control. The first practical commercial units were produced by Doug Spreng and Don Mathes as the Digicon and by Howard Bonner as the Digimite 8. Most American-made equipment uses amplitude modulation (AM). Frequency modulation (FM) is popular in some European countries, but is allowed in the U.S. only on the 53-MHz ham band. In both AM and

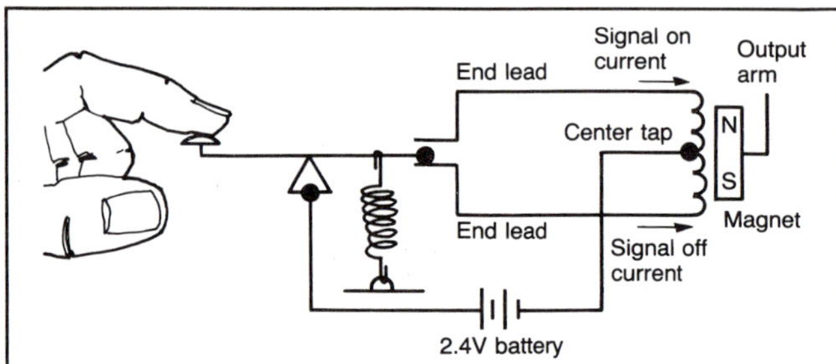

Fig. 2-2 PULSE SYSTEM PRINCIPLE

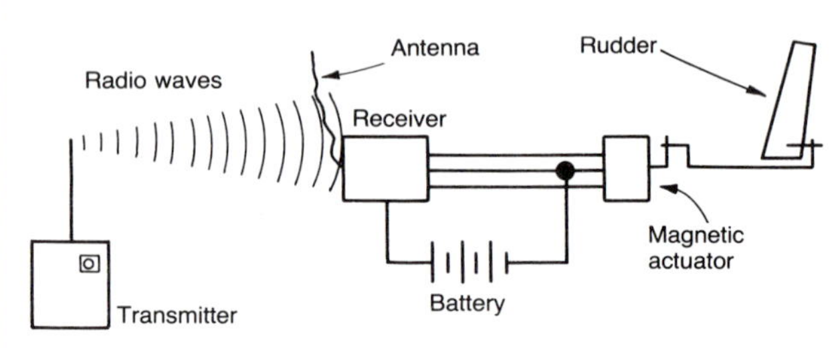

Fig. 2-3 PULSE SYSTEM BLOCK DIAGRAM

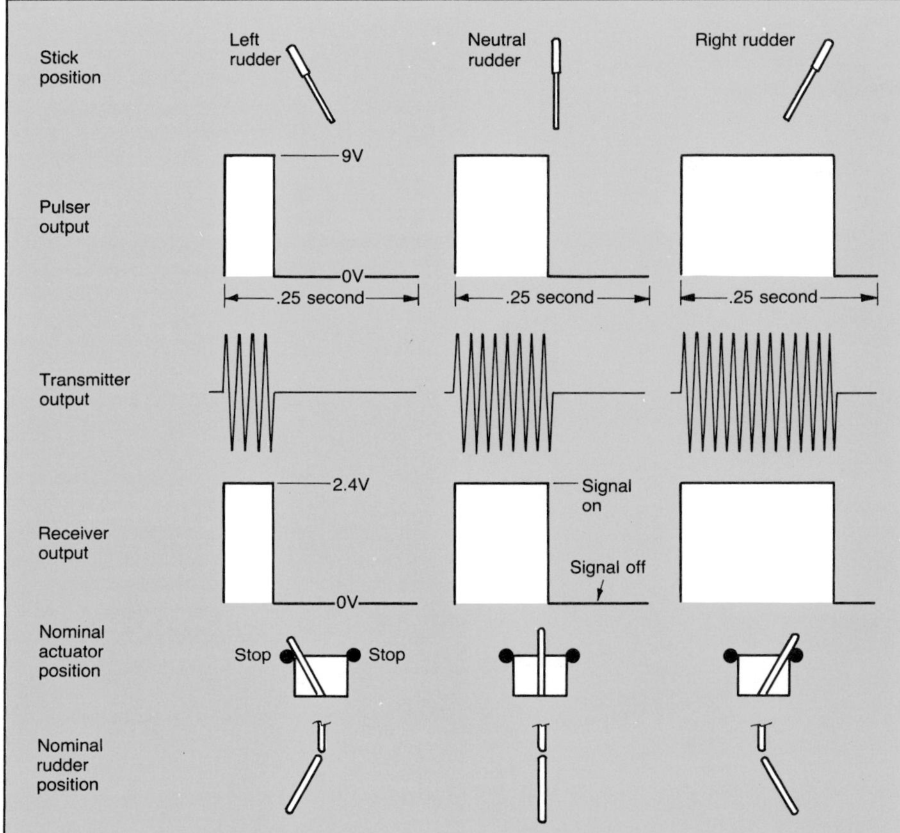

Fig. 2-4 HOW THE PULSER WORKS

FM, information is processed in packets of pauses and pulses known as frames. The modulation pattern is repetitious, occurring at a rate called the frame rate. Figure 2-6 is a simplified pictogram of one frame of information.

A system's frame is characterized by:

- The off time for the long portion of the frame known as the synchronization pause (sync pause).
- The number of channels.
- The time between pulses for all channels.
- The variation in the time between pulses for individual channels. *The control information is the relative length of time between control pulses for each channel.* Figure 2-7 shows how the control information is processed in a two-channel digital system. The principle is the same no matter how many channels are used.

Let's designate the length of the frame as TI, the synchronization pulse that starts each frame as T1, and the control pulses as T2, T3, and TN, where N represents the last channel. The difference between TI and T2 + T3 + TN is the synchronization pause. The frame rate may be fixed or variable. If the frame rate is fixed, TI is normally set by a master clock and the sync pause varies in length. The sync pause must always be long enough (usually about 5 milliseconds) for the decoder to reset with complete reliability. Most systems have a nominal, or neutral (control sticks and servos in their center positions), control pulse width of 1.5 ms with a variation of plus 0.5 ms and minus 0.5 ms for full control movements. Thus, in one frame the controls might all be positioned for the maximum pulse length of 2 ms. Using the two-channel example in Fig. 2-7, the channels would consume 4 ms at maximum width. If the sync pause is 6 ms, then the frame time, TI, is 10 ms. Frame rate is the inverse of frame time, so in this example the frame rate would be 100 frames per second. If both control pulses were short (1.0 ms), the sync pause would be 8 ms. In the same way, a seven-channel system would have a frame time of 7 times 2 ms plus 6 ms, or 20 ms, and the frame rate would be 50 per second — a more usual figure.

If the frame rate is variable, the system does not always have a master clock. If it does, the frame time is controlled by the channel encoders. A variable frame rate encoder maintains a fixed sync pause while the frame rate varies, the opposite of a fixed frame rate encoder. For the two-channel system example in Fig. 2-7, the sync pause would always be 6 ms but the frame time would range from 8 ms to 10 ms. They're the devil to track on an untriggered oscilloscope!

Few contemporary systems use variable frame rate because the servos operate away from the optimum frame time too often. In some older systems, this shows up as a change in resolution when servos move in one direction or the other.

Returning to Fig. 2-7, the encoder performs the frame sequence at the designated frame rate. A differentiator detects the fall of the encoded pulses as each channel sequences. The fall of waveform TI is also differentiated to produce the synchronization pulse that starts each frame. Thus, the encoder always produces N + 1 pulses where N is the number of channels.

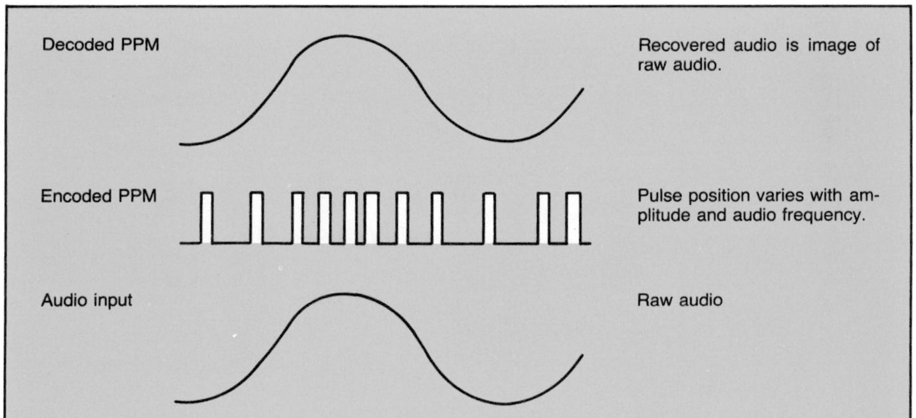

Fig. 2-5 PULSE POSITION MODULATION

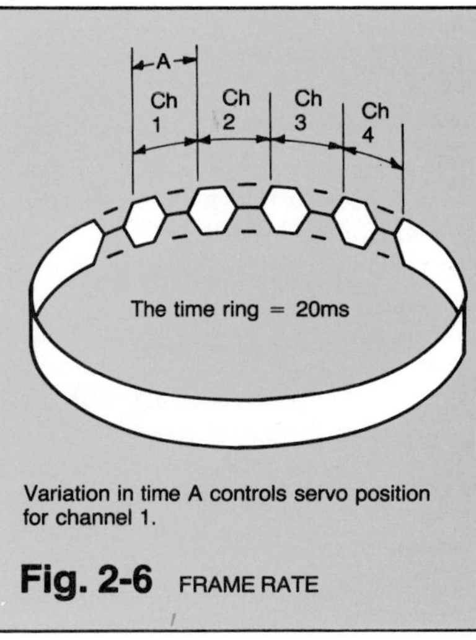

Variation in time A controls servo position for channel 1.

Fig. 2-6 FRAME RATE

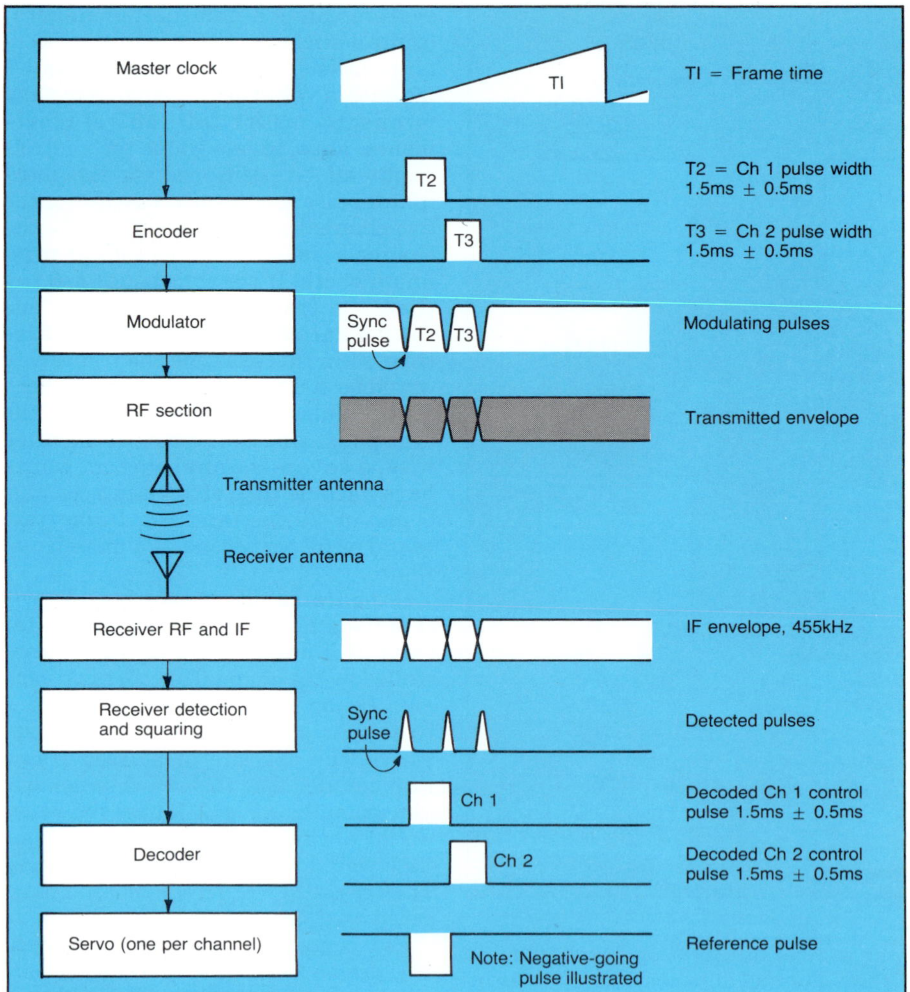

Fig. 2-7 DIGITAL SYSTEM

Modulator

The modulator accepts the output pulses from the encoder in one of many forms, for example, as a differentiated spike or a short square wave. It then performs a most important function insofar as FCC type acceptance is concerned: It converts each pulse to a precisely shaped modulation pulse. If the modulation pulse rose perpendicularly, the RF output would contain an infinite set of harmonics, a condition not only wasteful of power but also totally unacceptable to the FCC and modelers on adjacent frequencies. For both AM and FM, the modulation pulse must be shaped about like that in Fig. 2-8, usually with a base width of 250-350 microseconds.

The RF waveform in Fig. 2-8 is that used by almost all digital systems since 1964. The earliest digital sets used a form of modulation in which no carrier wave was transmitted except at the time of the modulation pulse. This proved disastrous because the receiver was open to interference during most of the frame — a situation that was corrected as quickly as possible. In recent years, one manufacturer, Westport International, has returned to that concept with a significant difference — they use sophisticated automatic gain control in their Variant systems to clamp the receiver during the off time. They chose this design to achieve greater power. The FCC limits average transmitted power to 5 watts on 27 MHz and 1 watt on 72 MHz. Average power is computed for the shaded area in the transmitted envelope, Fig. 2-7. Thus, 1 watt of power would have to be averaged over 10 ms minus about 1 ms for our hypothetical two-channel system, while the Variant design would be able to average 1 watt over just 1 ms. This is more than five times the peak power of conventional systems. However, this advantage is partly offset by the greater complexity of the receiver, which leads to more difficult troubleshooting. The FCC specifies the percentage of modulation precisely, and all systems must conform. Further, overmodulation does no good and produces frequency splatter.

FM systems use the same type of encoder as AM systems except that the modulator produces a shift in radio frequency during the modulation pulse. FM performance is specified by the deviation ratio, that is, by the ratio of the shift in frequency to the value of the nominal frequency. FM is further characterized as narrow band or broad band. FM for model use is narrow band because of FCC regulations and because we need many frequency slots. The permissible bandwidth is 5 kHz, resulting in a small deviation ratio at RC frequencies.

While narrow-band FM has the advantage of rejecting amplitude modulation caused by electrical noise in the model or from external sources, it does not provide significant rejection of unwanted AM RF signals. Thus we cannot take full advantage of the FM concept. To illustrate: The ultimate form of broad-band FM, used in sophisticated communications and radar systems, is called spread spectrum and the deviation ratio may be many orders of magnitude. These systems are virtually immune to jamming, even deliberate jamming. If the FCC would give us a 1-MHz bandwidth we could do grand things!

The RF section of the transmitter produces and amplifies a modulated RF waveform of the required frequency that has no more harmonics than the FCC allows. We want the minimum harmonic content practical at reasonable cost, Fig. 2-9, because the energy in the harmonics is wasted.

Receiver, decoder, and servos

The task of the receiver is to detect and reconstitute the transmitted modulation envelope under some of the most trying conditions imaginable. The receiver must perform equally well with the model three feet and 500 feet away, and at temperatures from 0 degrees F to 120 degrees F. The receiver and decoder are mounted on one PC board (single-deck receiver) or on separate PC boards connected via hookup wires (two-deck receiver). The detected pulses are fed to the decoder which synchronizes itself with the repeating sequence of each frame, sorts out the individual channel control pulses, and converts the channel pulses to square output pulses, Fig. 2-7. Each servo amplifier amplifies the decoder output for the channel to which it is assigned and uses it to command its servo's electric motor. Each electric motor is connected to a control surface via the servo gear train and the control linkages. That, in general terms, is how a digital RC system works.

Early encoders

A more complete understanding of the system's major units is helpful for those who plan to delve into kit-building, maintenance, and repair. Let's start with the encoder. It deserves treatment in some depth because it is the heart of the RC system. The first digital encoders used a closed-loop system called a ring counter that consisted

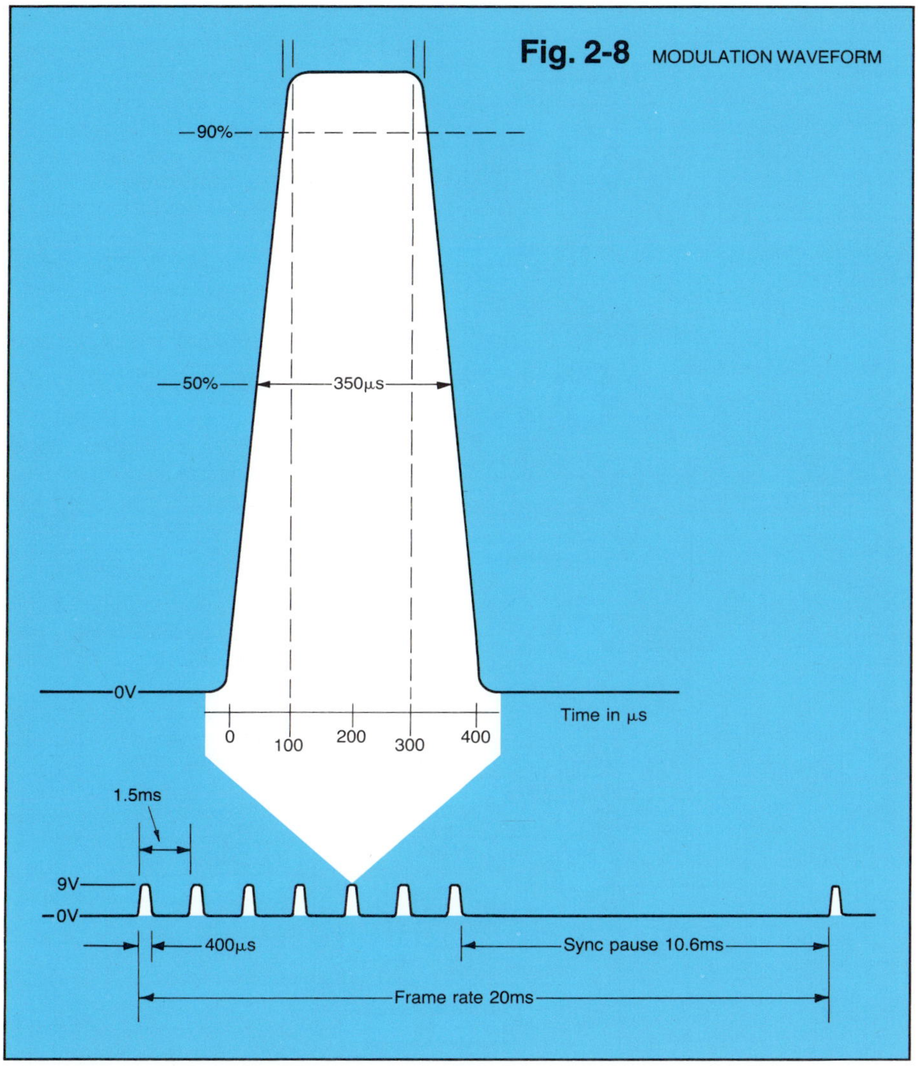

Fig. 2-8 MODULATION WAVEFORM

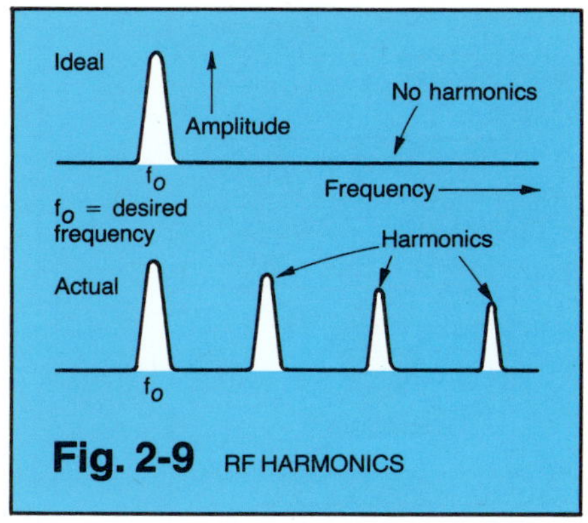

Fig. 2-9 RF HARMONICS

of a series of circuits called multivibrators. Multivibrators come in three varieties: the free-running (or bistable) multi, the one-shot (or monostable) multi, and a simpler variant, the half-shot multi. Once started, the free-running multi repeats its cycle continuously; its principal use is as a master clock. The one-shot or half-shot responds to a trigger pulse a few microseconds wide by producing an output pulse of a given pulse width. Such are the waveforms for T2 and T3, Fig. 2-7. Figure 2-10 presents typical multivibrator designs. The ring-counter encoder usually had enough unbalance somewhere in the circuit (or had a capacitor coupled to produce a spike somewhere in the circuit) to start the ring counter running. Once one of the multis was triggered, it cycled and started the next multi and so on. The sync pause was generated whenever one multi, set at a 5 ms time constant, was triggered. The control multis were set at 1.5 ms ± 0.5 ms by the control stick pots. Ring-counter encoders were soon abandoned because they often failed to start reliably.

Encoders of the 1960's and 1970's

The next development became almost standard: The general configuration was used with many refinements almost exclusively for over 10 years, Fig. 2-11. The master clock is a free-running multi formed by Q1, Q2, and related components. Notice that the free-running multi is assured of starting when capacitor C1 charges as SW-1 is turned on. The fall of the output from Q2 is coupled to the first control multi, a half-shot, formed by Q3 and related components, via the control pot R21. The falling edge of the square wave is differentiated (i.e., its slope is detected) by C21 and R20. The positive spike is coupled to the modulator to form the first, or sync, pulse. The same action across C30 triggers the aileron one-shot. Q3 stays off until C30 discharges the nominal 1.5 ms ± 0.5 ms through control pot R21. When Q3 completes its cycle, the differentiation action of C31 and R30 couples the next spike to the modulator via D30; C40 charges and the next one-shot cycles. The process continues until all channels have cycled. The encoder then pauses until the clock starts another

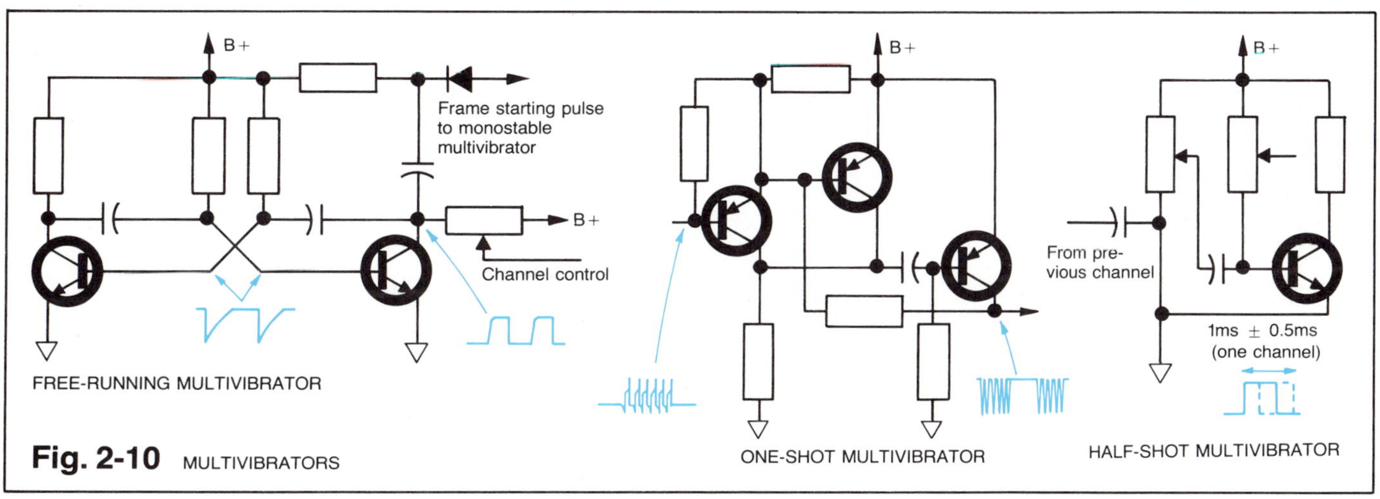

Fig. 2-10 MULTIVIBRATORS

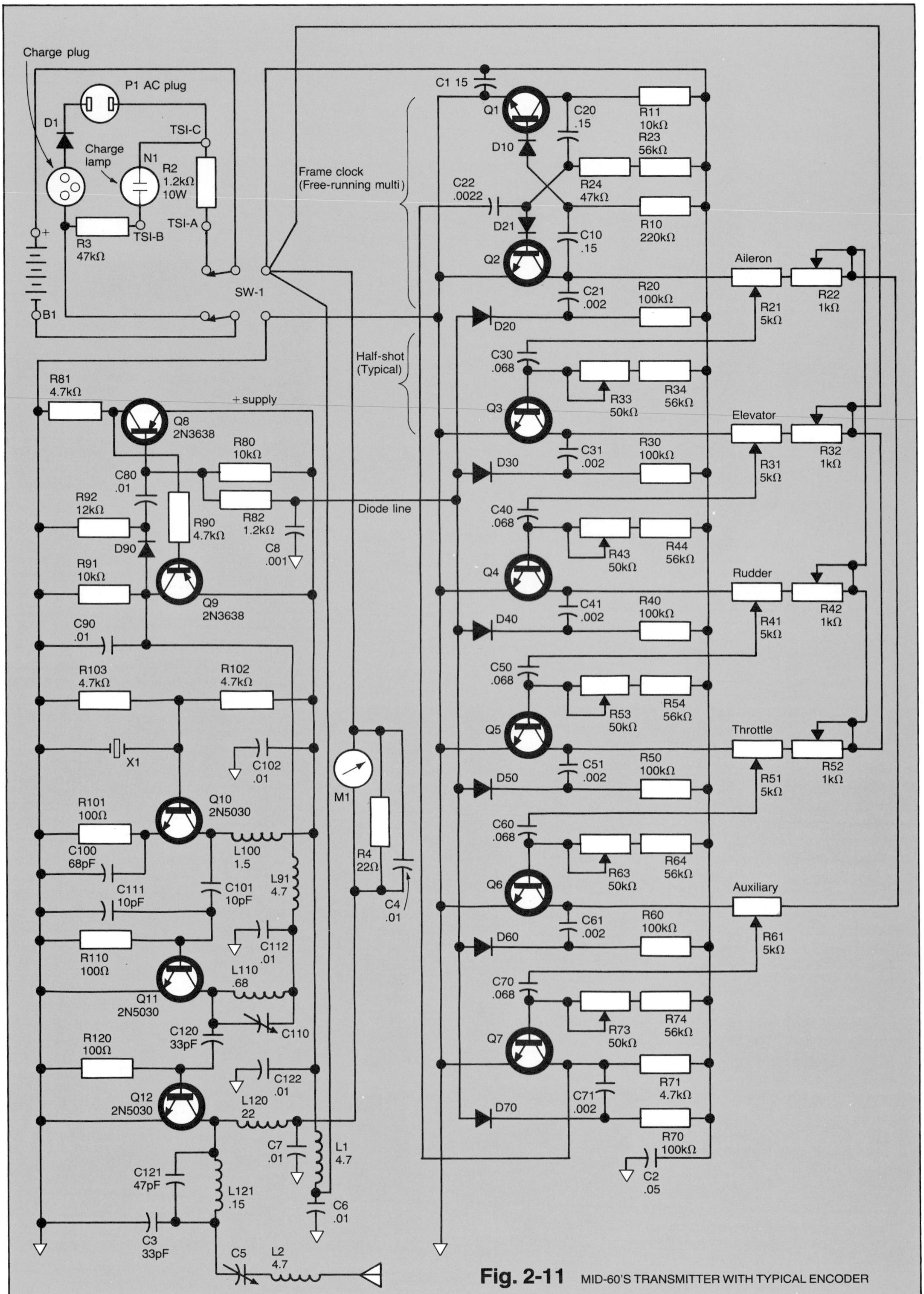

Fig. 2-11 MID-60'S TRANSMITTER WITH TYPICAL ENCODER

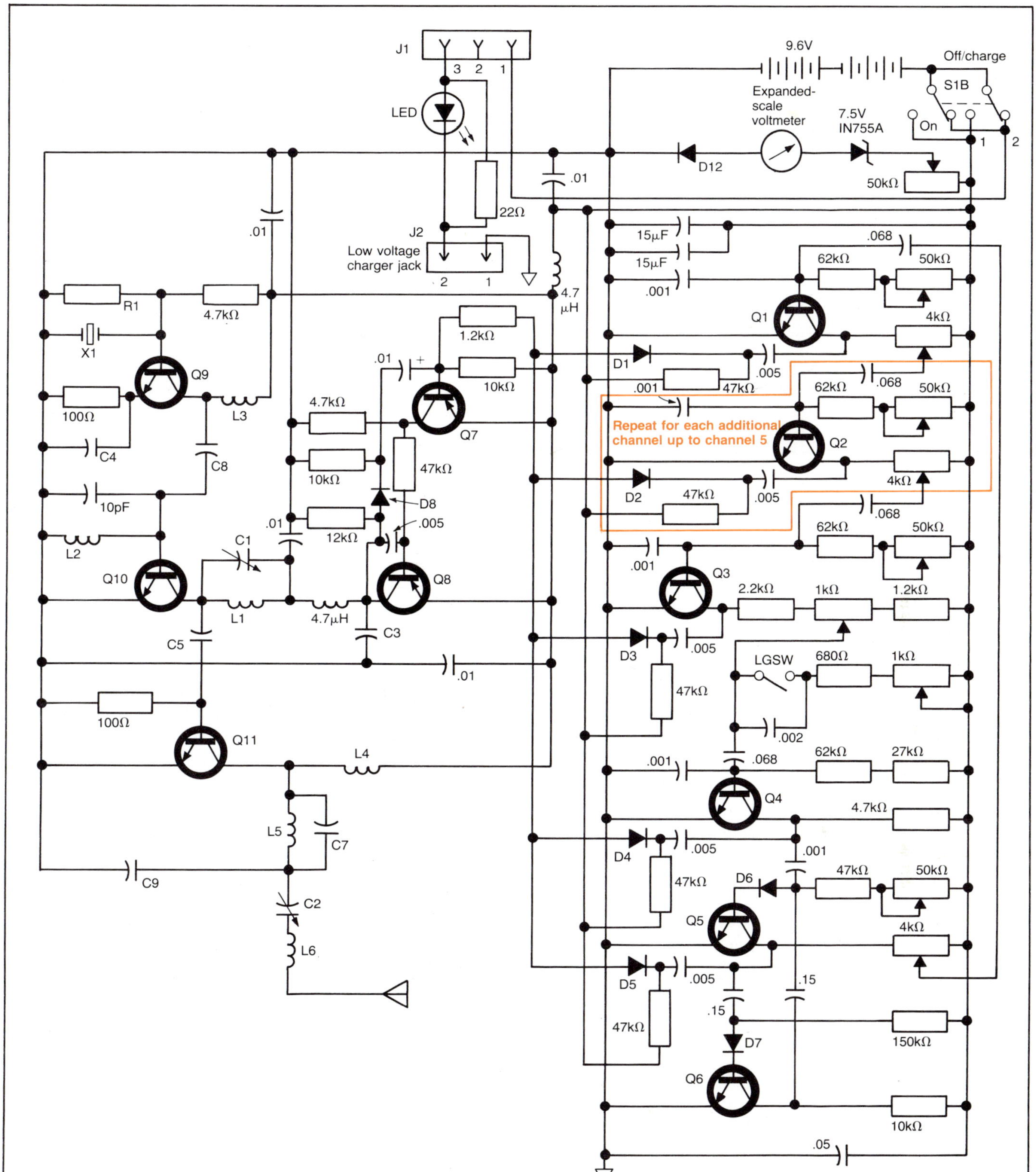

Fig. 2-13 MID-70'S TRANSMITTER WITH TYPICAL ENCODER

frame. All this happens at up to 50 times per second. The master clock function could just as well have been performed by a relaxation oscillator similar to that formed with a unijunction transistor (UJT) in Fig. 2-12. The circuit in Fig. 2-11 is for a transmitter designed in the mid-60's, a good, solid design. The circuit in Fig. 2-13 is for a 1976 design by EK International; study will show many refinements but also many similarities.

The differentiated spikes appearing on the diode line in Fig. 2-11 are not suitable for RF deck modulation. Therefore, most sets use a one-shot multi to square and set the length of the modulation pulse — the function performed by Q8, Q9, and related components in Fig. 2-11. C90 shapes the pulse to match that of Fig. 2-8. This is now usually performed on the encoder board when separate RF deck and encoder PC boards are used. In this way, the RF deck can be a separate module to allow for frequency change with a minimum of parts.

At various times, Ace, Micro, World Engines, and others used resistor-transistor logic (RTL) or transistor-transistor logic (TTL) integrated circuits in encoders, but these usually employed gates on inverters to form multis.

IC encoders

In the mid-70's, the first integrated circuits at the medium-scale integration (MSI) level became available. MSI incorporates many circuits on one IC chip. For example, an eight-bit shift register IC with eight one-shot multis and lots of other circuitry performs steps that used to require a whole PC board of transistors and other components. Figure 2-14 presents an IC encoder I designed for Ace R/C. It uses two MSI ICs — a binary-coded digital counter and a decoder driver to perform a sequential stepping function known as commutation. Similar designs by Kraft (Fig. 2-15), Pro-Line, and RS have used shift registers to commutate. Exclusive of the control pots, the complete Ace commutated encoder, including voltage regulator and power supply filter, consists of just 25 components. Compare this with the far greater number of components in earlier encoders, Figs. 2-11 and 2-13. The commutated encoder needs only one timing circuit for all channels instead of one timing circuit for each channel required by previous encoders. The control input to the timing circuit steps sequentially from one control potentiometer to the next until a complete frame of information is encoded. This function of stepping sequentially is termed commutation. The Ace encoder consists of a free-running multivibrator that performs the timing function, a control-input buffer for this multi, and two ICs that commutate the multivibrator time constant from one control input to another.

The free-running multi consists of Q2 and Q3, accompanying resistors R1 through R8, and capacitors C1 through C4. I'll describe the period when the collector of Q2 is high, that is, nearly 9.6 volts. This occurs when Q2 is biased off so that its base is near ground voltage. The multi changes state in a clocklike manner, switching at a regular interval set by the value of the related resistors and capacitors. Its on-time may equal off-time, or these times may be independently varied. In this case, I've set the off-time at 250 microseconds by choosing the values shown for C2 and R5. This period is the nominal width of the modulation pulse.

On-time, set by the value of C1, R4, and the voltage level at the emitter of Q1, is variable by ± 0.5 ms about a nominal time of 1.5 ms. It can be shortened to 1.0 ms and lengthened to 2.0 ms and is continuously variable between these extremes. Q1, an NPN, turns on when its emitter goes more positive, lengthening the on-time. When its emitter goes more negative, the on-time shortens. The variation in the bias level for Q1 is controlled sequentially by each of the control potentiometers. Note that the retract function is just another form of this arrangement. It allows a switched pulse of 1.0 ms or 2.0 ms for two-position functions such as retractable landing gear or flaps.

IC-1 and IC-2 provide the bias required to achieve the control time periods of 1.5 ms ± 0.5 ms and the bias required to produce a sync pause of about 4.5 ms between each frame. IC-1 is a decade counter that converts decimal inputs to binary-coded outputs. The logic code, Fig. 2-16, shows that it accepts sequential clock pulses and produces binary-coded outputs at A, B, C, and D. This is the same as counting decimally except that our friend can only count to one. Thus the A output goes high (1) on the first clock pulse, then low (0) on the second. This is the same as counting to nine and then making the last significant place zero while the second becomes a one to form ten. When A goes low it carries to the next significant place just as in going from nine to ten and the B output goes high. It cannot go low again until A goes low once more. The same relationship is true for B versus C and for C versus D. In this way, A changes state with every count, B with every other count, C with every fourth count, and D with every eighth count. Similar counters are used in all digital display systems.

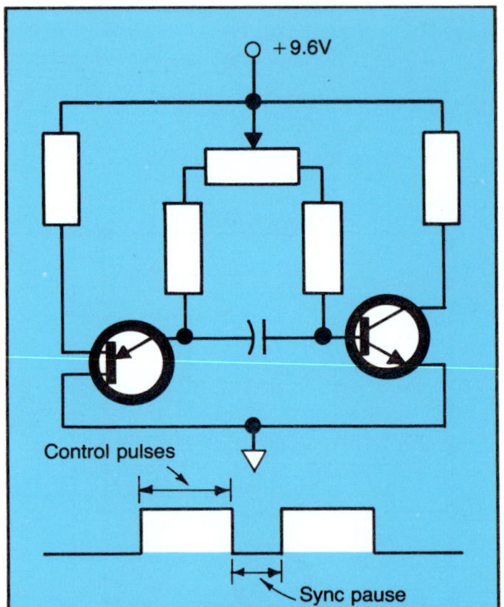

Fig. 2-12 UNIJUNCTION TRANSISTOR CLOCK

IC-2 is a binary-coded decimal to de-

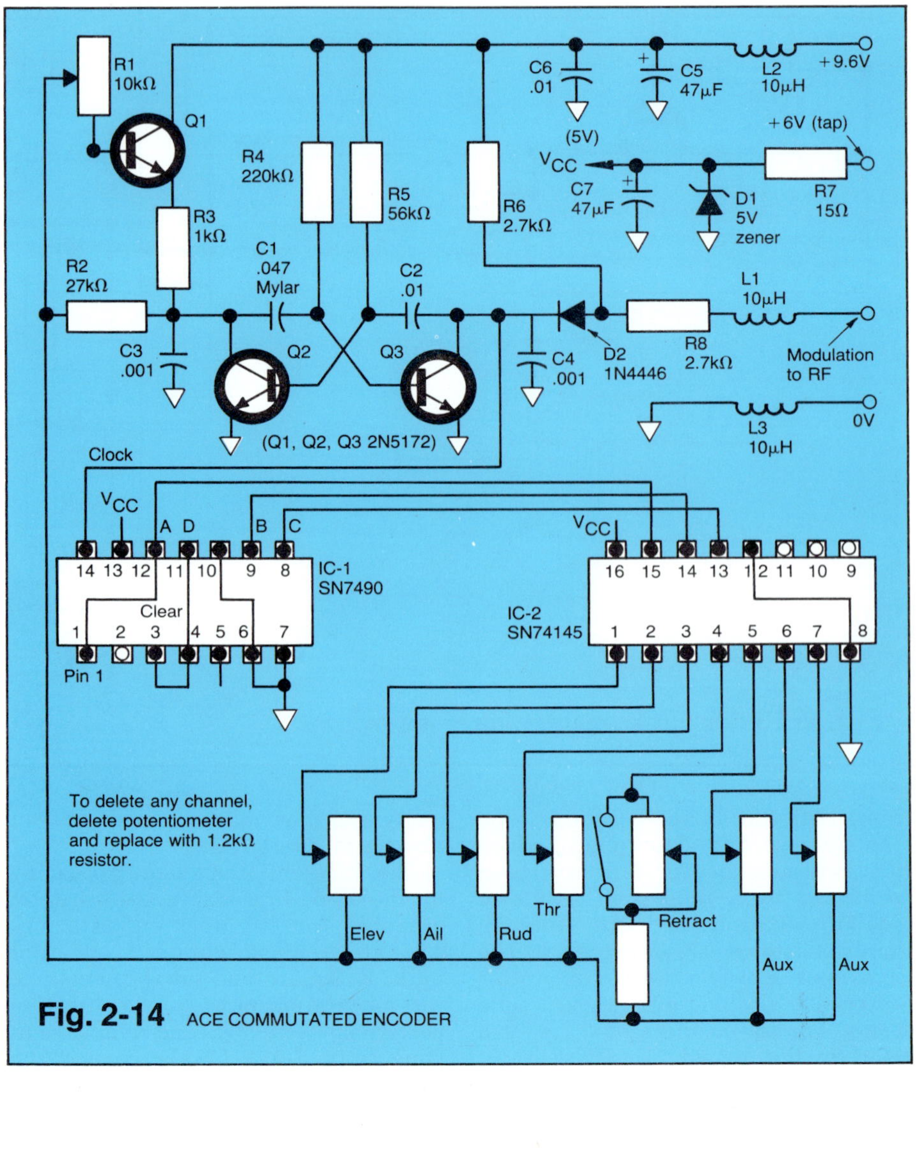

Fig. 2-14 ACE COMMUTATED ENCODER

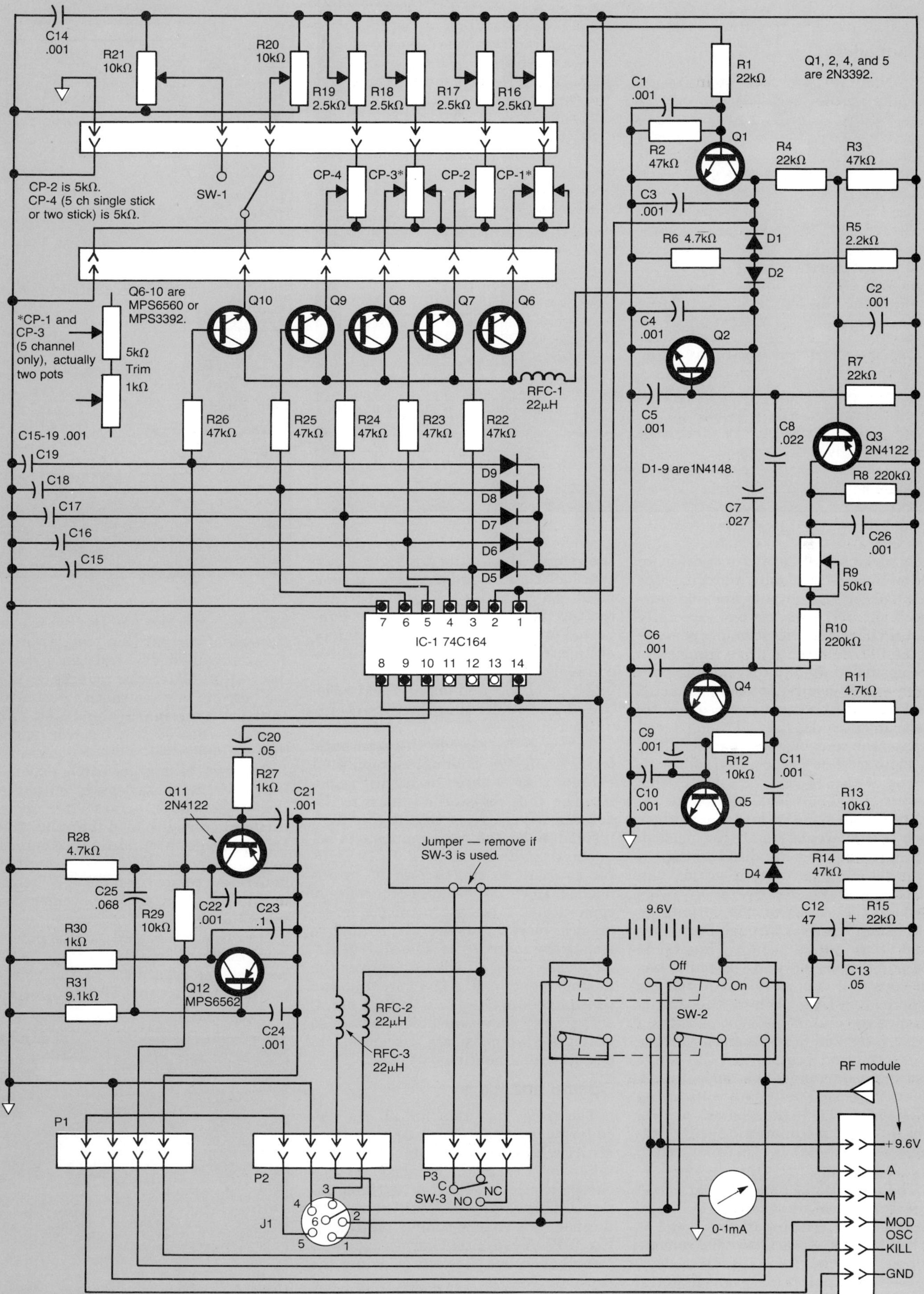

Fig. 2-15 KRAFT TRANSMITTER WITH COMMUTATED ENCODER

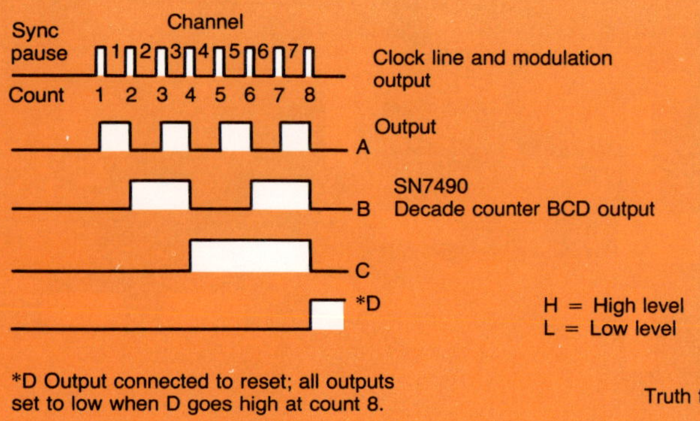

Count	Output			
	D	C	B	A
0	L	L	L	L
1	L	L	L	H
2	L	L	H	L
3	L	L	H	H
4	L	H	L	L
5	L	H	L	H
6	L	H	H	L
7	L	H	H	H
8	H	L	L	L
9	H	L	L	H

H = High level
L = Low level

Truth table SN7490 Decade counter

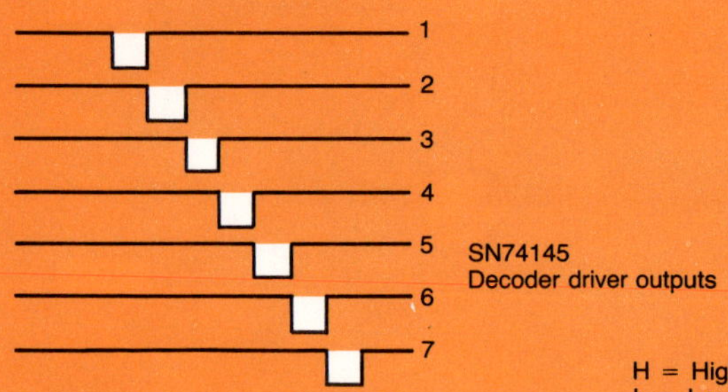

Input				Output low
D	C	B	A	
L	L	L	L	O
L	L	L	H	1
L	L	H	L	2
L	L	H	H	3
L	H	L	L	4
L	H	L	H	5
L	H	H	L	6
L	H	H	H	7
H	L	L	L	8
H	L	L	H	9

H = High Level
L = Low level

Truth table SN74145 BCD to decoder driver

Fig. 2-16 LOGIC CODES

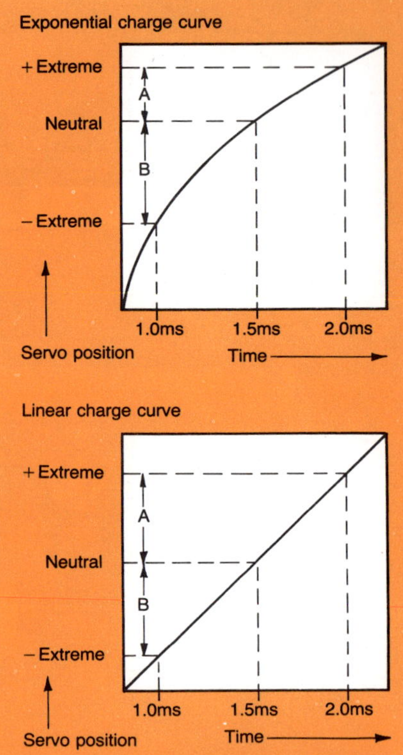

Fig. 2-17 LINEAR AND EXPONENTIAL CHARGE CURVES

coder driver. Its usual function is to accept binary-coded data and convert it to a one of ten output. For our purposes, the SN74145 has one especially useful characteristic: It is an open-collector TTL device. Its drive transistors are normally used to drive display devices and it has no internal collector load. This is important because it automatically provides the isolation needed to keep the control pots separated from each other when the outputs (1 through 7, and 9 through 11) are high. This means that the blocking diodes required with other counting devices such as shift registers are not needed. These ICs do have the disadvantage of somewhat higher current drain than shift registers. As SN74145 steps through the sequence, the outputs go low. When inputs A through D are low, then all outputs are high. Thus Q1 is biased strongly on and the multi has the long (4.5 ms) period required for synchronization. When the multi reaches the end of the sync pause and goes off for 250 microseconds, a frame of information begins, i.e., the multi issues the start pulse that appears as a 250-microsecond pulse of 9 volts at the collector of Q3. It then passes via D2 and R8 to the modulator in the RF section. Notice that it is also connected to the clock input of IC-1 at pin 14.

The SN7490 is able to clear itself. The D output (pin 11 of IC-1) is connected to the clear input (pins 3 and 4). When the count reaches eight (remember we need one more pulse than channels — eight pulses for seven channels) and D starts to go high, IC-1 is cleared; all outputs go low. The code for IC-2 shows that all outputs are high; a new sync pause starts and the frame is complete. We could use the same approach to stop the count at two (reset connected to B) to provide one channel, or at four (reset connected to C) for three channels. However, it is simpler to delete control pots and install fixed resistors to provide blank outputs for unused channels.

Figure 2-15 presents the schematic for a Kraft five-channel commutated encoder that is based on a shift register. The shift register is similar to the decoder driver except that it goes high instead of low as each segment is sequenced. It does not need BCD input for control, so the second IC is not needed. However, because plus voltage commutates to each control pot in sequence, blocking diodes are needed to isolate the control pots. The shift-register encoder has an advantage over the arrangement in Fig. 2-14 in that the control pots are used as voltage dividers rather than as variable resistors so that the system is less susceptible to pot noise and nonlinearity.

Opamp encoders

The time constants for all the encoders discussed so far are set by a multi which depends on the time it takes for a capacitor to charge or discharge through a fixed or variable resistor. This creates difficulties because a capacitor's charge time is not linear, Fig. 2-17. A capacitor charges exponentionally because the voltage across the capacitor rises as it charges via a fixed resistance from a fixed voltage. Thus, the voltage differential across the resistor becomes less and less. Ohm's law states that current flow equals voltage difference divided by resistance, and as the voltage difference reduces, so does the current flow, so that the capacitor charges ever more slowly as time passes — theoretically it never reaches full voltage.

Because the charge rate is exponential, and the control pulse output reflects that curve, servos often moved farther in one direction than the other. Designers got around the problem because the servo also used a multi to generate a comparison pulse. When set up in opposition, servo movement was more linear.

Now that it's been done, the obvious solution is to force the current flow to be constant regardless of the voltage across the capacitor. This became eco-

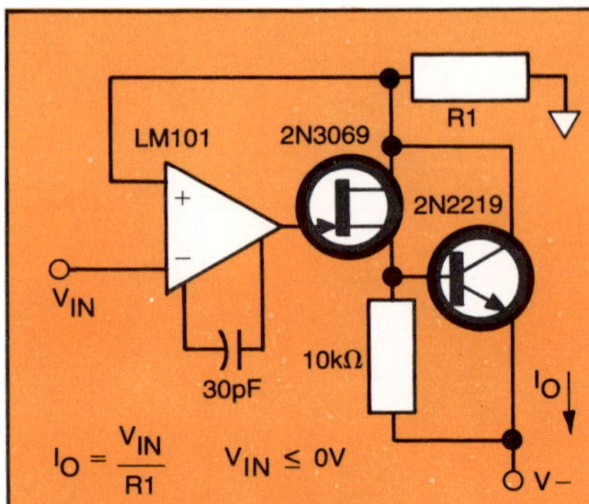

Fig. 2-18 PRECISION CURRENT SOURCE

Fig. 2-19 KRAFT TRANSMITTER WITH LINEAR-RAMP ENCODER

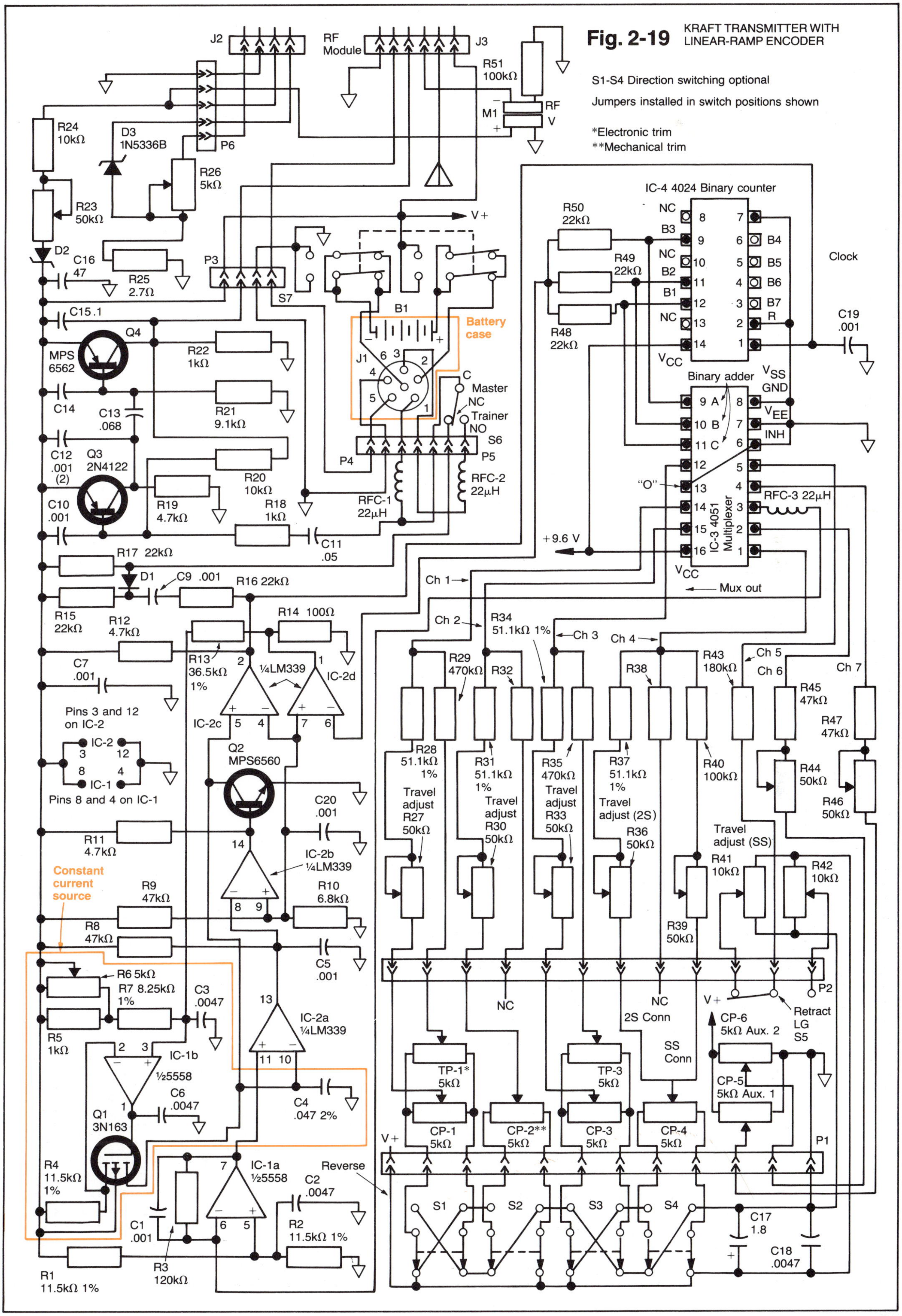

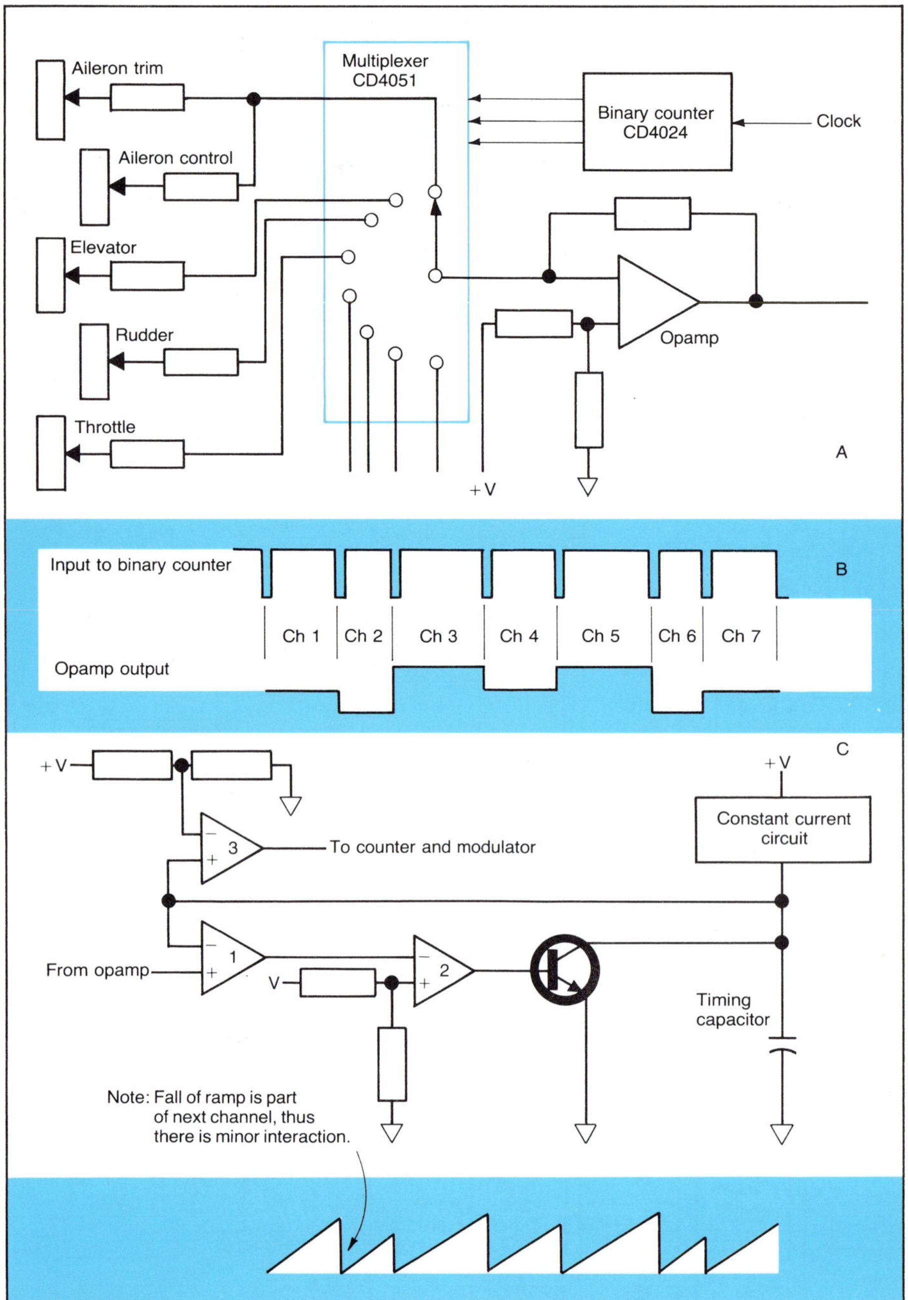

Fig. 2-20 KRAFT LINEAR-RAMP ENCODER PRINCIPLE

nomically practical when low-cost versions of a most modern IC called the operational amplifier (opamp) appeared. The opamp has tremendous sensitivity to minute voltage differences at its control inputs. Such a circuit, Fig. 2-18, was published in a National Semiconductor applications note several years ago and I adapted it. In that circuit V_{IN} is a fixed reference voltage. Any minute difference between V_{IN} and the positive input voltage to LM101 appears magnified at the pointy end that controls the bias on 2N3069 and thus 2N2219. As the capacitor (theoretically connected to the emitter of 2N2219) charges, the current across R1 will be I_O. As capacitor voltage rises (the bucket fills), I_O would normally start to decrease. However, LM101 senses the voltage drop across R1, calculated from Ohm's law as voltage difference equals current flow times resistance, and adjusts to maintain constant current flow. Now our capacitor charges linearly and we have equal servo travel.

Doug Spreng, then at Kraft, took advantage of this principle about 1974. It might seem logical that it could be put on any of the earlier designs but that isn't practical because we would need one such unit per channel. The basic encoder that Doug designed used the circuit in Fig. 2-19 in the Kraft Signature Series systems. I call it a linear-ramp encoder. It is a commutated encoder that took advantage of a then-new device called the analog switch. It passes whatever voltage appears at the input (unlike the shift register or decoder driver) through each of eight solid state switches in sequence. The 4051 analog switch requires BCD input for control of its sequence, as did the SN74145. It sequences between the seven inputs internally to produce a commutated output at pin 3. It doesn't need blocking diodes.

Jim Oddino has provided a clear description of the Kraft encoder's design: "This discussion will show how the encoder accepts inputs from a number of channels and feeds them into one opamp and then how the output of the opamp is transformed into a pulse width.

"The device that allows us to share one opamp with all channels is called a multiplexer. We can think of a multiplexer as an eight-position switch, Fig. 2-20a.

"If we picture the switch being stepped at the end of each timing interval, we can see that the output of the opamp will be controlled by the voltages set on the control pots (and other functions that might be summed in). The output of the opamp might look like Fig. 2-20b.

"Notice that the top trace looks just like the thing we need to modulate our RF section. If we can convert the voltage out of the opamp to this waveform, we'll be in business.

"A straightforward means of generating the familiar ramp waveform is by making an oscillator with a couple of comparators, a transistor switch, and a timing capacitor, Fig. 2-20c.

"The timing capacitor charges linearly until it reaches the voltage out of the opamp. At that point, comparator 1 switches from high to low, comparator 2 inverts this and turns on the transistor, discharging the capacitor, at which time it starts charging again. Whenever comparator 1 switches, 3 also switches and produces the desired waveform to modulate the RF and clock the multiplexer."

The NE5044 encoder

In early 1979, Signetics introduced an integrated circuit that performs the functions of the most complex linear-ramp encoder. The NE5044, Fig. 4-1, includes a multiplexer switch, constant-current device, master clock, modulation output one-shot, and throws in a voltage regulator to boot. It is a sophisticated device that produces a linear sawtooth that ramps both up and down. All input potentiometers are set up as precision voltage dividers between V_{REF} and ground. The NE5044 has a high impedance so that when the inputs are switched off the input resistance is several million ohms. When the inputs are switched on the resistance is nearly zero. This feature, plus the use of precision divider input pots, allows the inputs to be manipulated in many ways for just about any function imaginable. It's likely that the NE5044, and

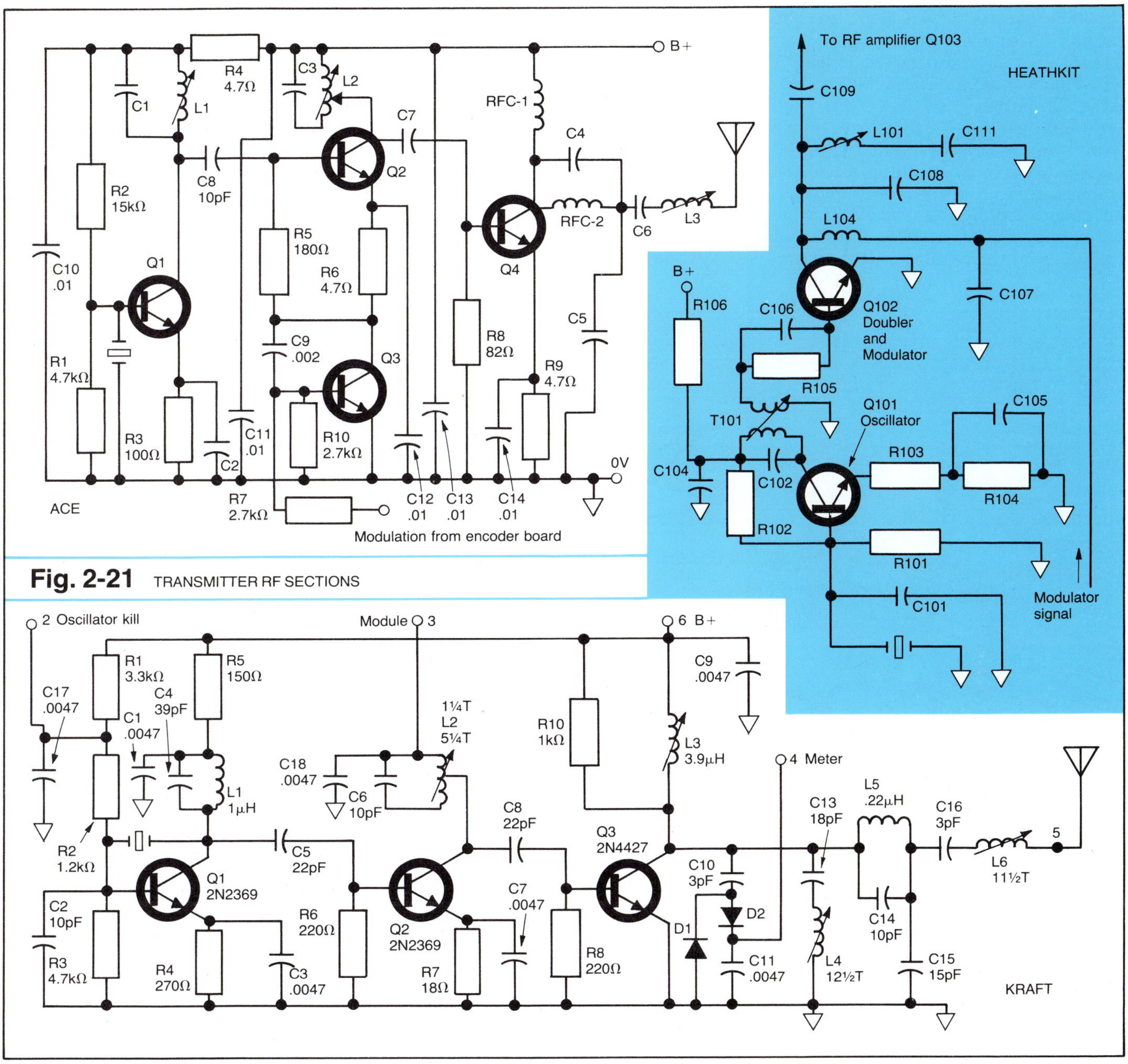

Fig. 2-21 TRANSMITTER RF SECTIONS

possibly similar devices by other manufacturers, will become standard throughout the industry in the early 1980's. The NE5044 is described in more detail in Chapter 4.

Transmitter RF sections

The RF section is as simple as the encoder is complex. Figure 2-21 shows three RF decks. Most RF sections consist of two or three parts — an oscillator, a frequency doubler (if used), and a power amplifier. All RC transmitters are crystal controlled, which means that a quartz crystal, ground to extremely fine tolerances, determines the transmitter output frequency, provided that all circuits are properly tuned. The oscillator is an amplifier with some of its output fed back to its input. Therefore, the crystal, which acts as a sharply tuned circuit, oscillates, producing RF output at exactly the desired frequency or at exactly half the desired frequency if a doubler is used. A tuned circuit couples the oscillator output to the power amplifier. If a frequency doubler is used, the oscillator output passes through the doubler first. The PA is simply an amplifier; it shouldn't oscillate. Poor design or faulty components can cause it to do so — usually on a frequency we don't want. The PA should duplicate the signal it receives from the oscillator or doubler, but treble or quadruple its power by drawing current from the transmitter battery pack. Most transmitters use a tuned circuit to match the PA transistor output circuit to the antenna (called impedance matching). They would match poorly otherwise, and power would be wasted by being turned into heat in the transistor. Some RC manufacturers warn against extensive testing with the transmitter antenna collapsed, as it would be for short-range checks. Collapsing the antenna detunes the output circuit, and may overheat the transistor.

More about RF sections

In the Ace pulse proportional transmitter the modulator impresses an audio frequency tone on the RF output. The tone comes from a separate AF oscillator. This tone may be fed to the oscillator, to the PA, to both, or to an interim stage called a buffer. In any case, the result is to turn the PA on and off smoothly at an audio frequency tone that produces, ideally, a sine wave. You might wonder how this is possible if the modulator is connected only to the oscillator. It is possible because transistor PAs have the unique property of drawing very low current — often only a few milliamps — unless they are driven by RF from the oscillator. The oscillator, in turn, can be turned on and off at an audio frequency by the modulator and the PA will follow in step to produce a modulated RF output.

Some digital RF units have more than two transistors. For example, many 72-MHz units have an extra transistor stage, a frequency doubler, between the oscillator and PA. As the name suggests, it doubles the output frequency of the oscillator. It is an amplifier tuned to the second harmonic of the input frequency so that its output is twice the frequency of the input. The crystal in such a transmitter oscillates at half of the actual output frequency. Doublers are used because lower frequency crystals are less expensive, less fragile, and more efficient than those for higher frequencies.

The tuned circuit between the PA and the antenna, as noted above, serves mainly to ensure efficient coupling, but it also helps to suppress undesired harmonics. Inferior amplifiers produce not only an amplified copy of the desired frequency, but also second, third, and other harmonics, Fig. 2-9. Thus a 27-MHz transmitter can put out a strong signal on 54 MHz, on 81 MHz, and so on. Proper amplifier design and tuning circuits between the PA output and antenna can reduce these harmonics to a low level, ensuring that little power is wasted.

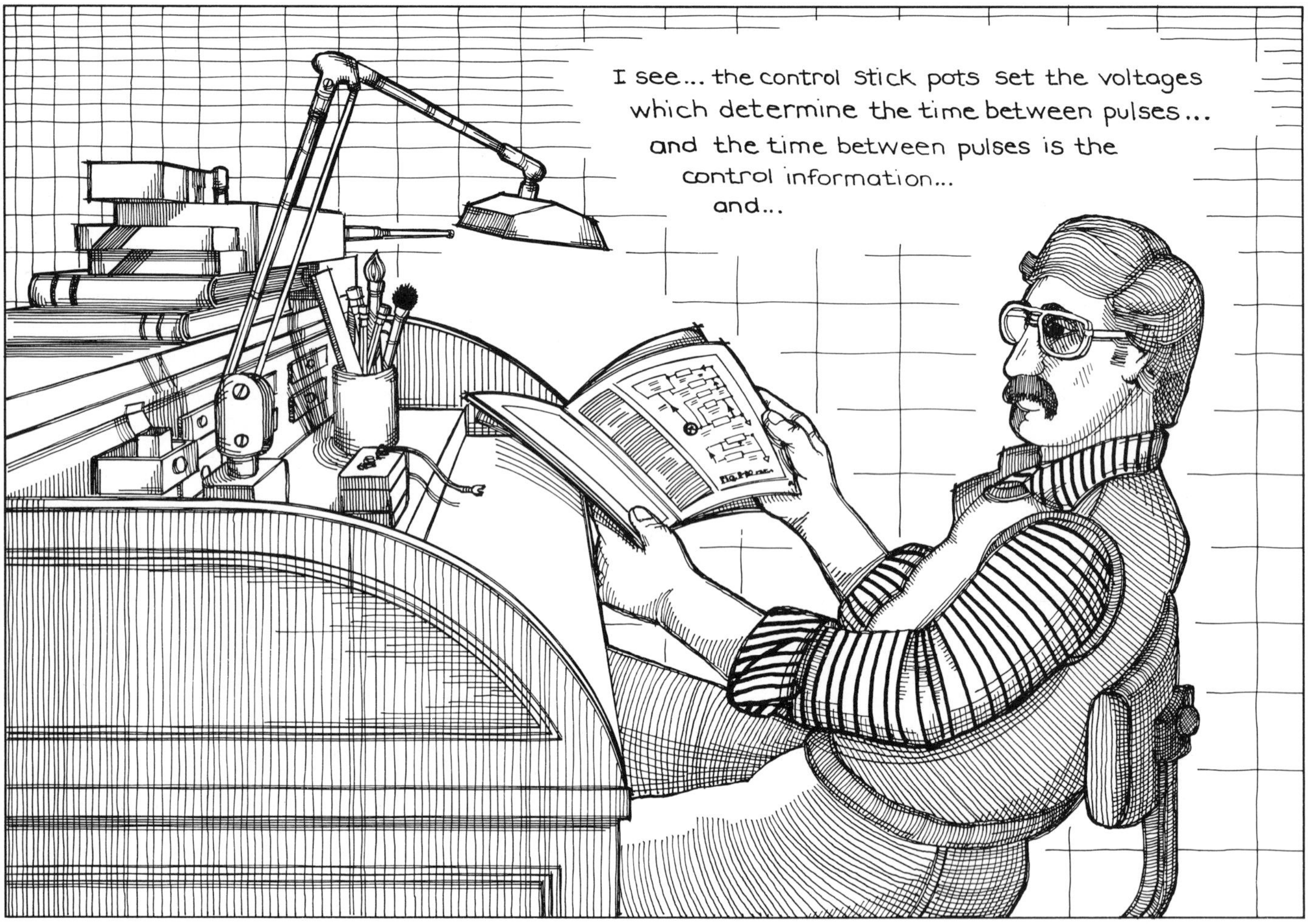

3 Receiver and servo design

The receiver has undergone intense design activity over the years. The reasons lie in:

- The need for better selectivity to avoid interference and to permit more frequency slots.
- The requirement for better noise rejection made necessary by powerful, fast-switching servo amplifiers that put voltage spikes on the line.
- The market's demand for smaller size, lighter weight, and greater reliability. Even so, receivers have much in common: All are superheterodyne, almost all use double RF tuning, almost all operate at an intermediate frequency of 455 kHz, and all AM receivers have automatic gain control. All FM receivers have limiting.

Let's examine how a superheterodyne receiver works in broad and nontechnical terms. Figure 3-1 shows the major components. All the blocks have battery power, of course, but I've eliminated the power supply leads for simplicity. Starting with the antenna at far left, the incoming signal passes to L1. If the receiver has a double-tuned input circuit, L1 is coupled directly to L2 by capacitor C1 as shown, or by other coupling means. In some receivers, a transistor couples the two inductors. This boosts the signal but isn't absolutely necessary. Some superhets have only a single input inductor; the antenna is connected directly to L2, and L1 is omitted. Two inductors, however, with or without an RF transistor between them, give much sharper input tuning and better image rejection.

The first box is the mixer, so called because it mixes the incoming signal with the signal from the crystal-controlled high-frequency local oscillator directly below it. The mixer also contributes some amplification. The oscillator in a superhet on the lower RC frequencies is tuned to operate at a frequency differing from the signal frequency by exactly the intermediate frequency of the receiver. The intermediate frequency is almost always 455 kHz. The oscillator frequency may be higher or lower than the signal frequency. For example, for a signal frequency of 26.995 MHz, the local oscillator frequency is 26.540 MHz or 27.450 MHz. Either crystal frequency produces a difference of 455 kHz when mixed with the 26.995-MHz signal frequency. The 455-kHz intermediate frequency is then amplified in the IF section of the receiver. Note that with the lower crystal frequency, a signal on 26.085 MHz would also produce a difference of 455 kHz and could pass through the IF section.

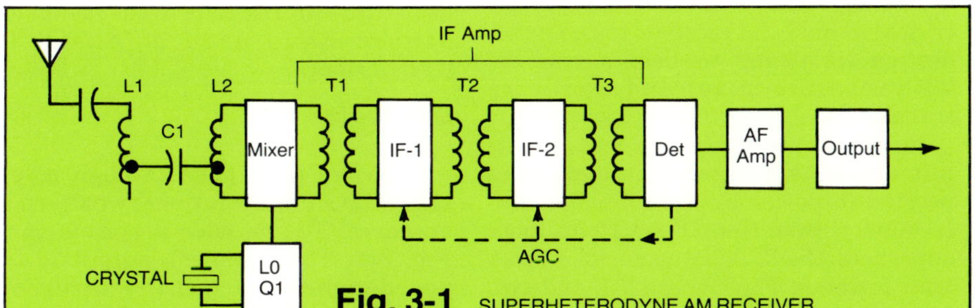

Fig. 3-1 SUPERHETERODYNE AM RECEIVER

Crystal Frequency		Operating Frequency	Flag Colors
35.8125 MHz x 2 = 71.6250 MHz	+ 455 kHz =	*72.0800* MHz	Brown and White
35.8525 MHz x 2 = 71.7050 MHz	+ 455 kHz =	*72.1600* MHz	Blue and White
35.8925 MHz x 2 = 71.7850 MHz	+ 455 kHz =	*72.2400* MHz	Red and White
35.9325 MHz x 2 = 71.8650 MHz	+ 455 kHz =	*72.3200* MHz	Violet and White
35.9725 MHz x 2 = 71.9450 MHz	+ 455 kHz =	*72.4000* MHz	Orange and White
36.2525 MHz x 2 = 72.5050 MHz	+ 455 kHz =	*72.9600* MHz	Yellow and White
37.5925 MHz x 2 = 75.1850 MHz	+ 455 kHz =	*75.6400* MHz	Green and White

Table 3-1 72 MHz CRYSTAL FREQUENCIES

Doubler crystals

Most receivers on 72 MHz use a doubler crystal, that is, the local oscillator operates at half the desired frequency and the mixer is driven by the second harmonic via a tuned circuit. Doubler crystals are used in receivers on the higher frequencies for the same reasons they are used in transmitters: lower frequency crystals are stronger and cheaper. Table 3-1 lists doubler crystal frequencies. The italicized numbers are those you'll hear at the flying field. The crystal frequencies shown are for the local oscillator second harmonic 455 kHz below the channel frequency. If the system is designed for 455 kHz above the channel frequency, the crystal frequency is the channel frequency minus 455 kHz divided by two.

Spurious response

Undesired signals, called images, at exactly twice the intermediate frequency, 910 kHz, away from the signal frequency can cause trouble on all of the RC frequency bands. For example, if the receiver is tuned to 26.995 MHz, strong signals on 26.085 MHz or 27.905 MHz could bother the receiver, depending upon whether the receiver oscillator is tuned lower or higher than the signal frequency. Therefore, if mysterious interference develops on any RC spot, first find out if the receiver oscillator is lower or higher than the signal frequency. Then check for strong signals at twice the IF away. On the 27-MHz RC band, the interference is likely to come from a transmitter in the citizens band. This is not true on the 53-MHz and 72-MHz bands, however. Such image interference has sometimes been the cause of a problem when a modeler was certain there was no interfering signal on his transmitter frequency. It is well to be aware of this potential problem.

Oscillators and detectors

Some further points on receivers should be mentioned. First, why is 455 kHz such a magic number that it is

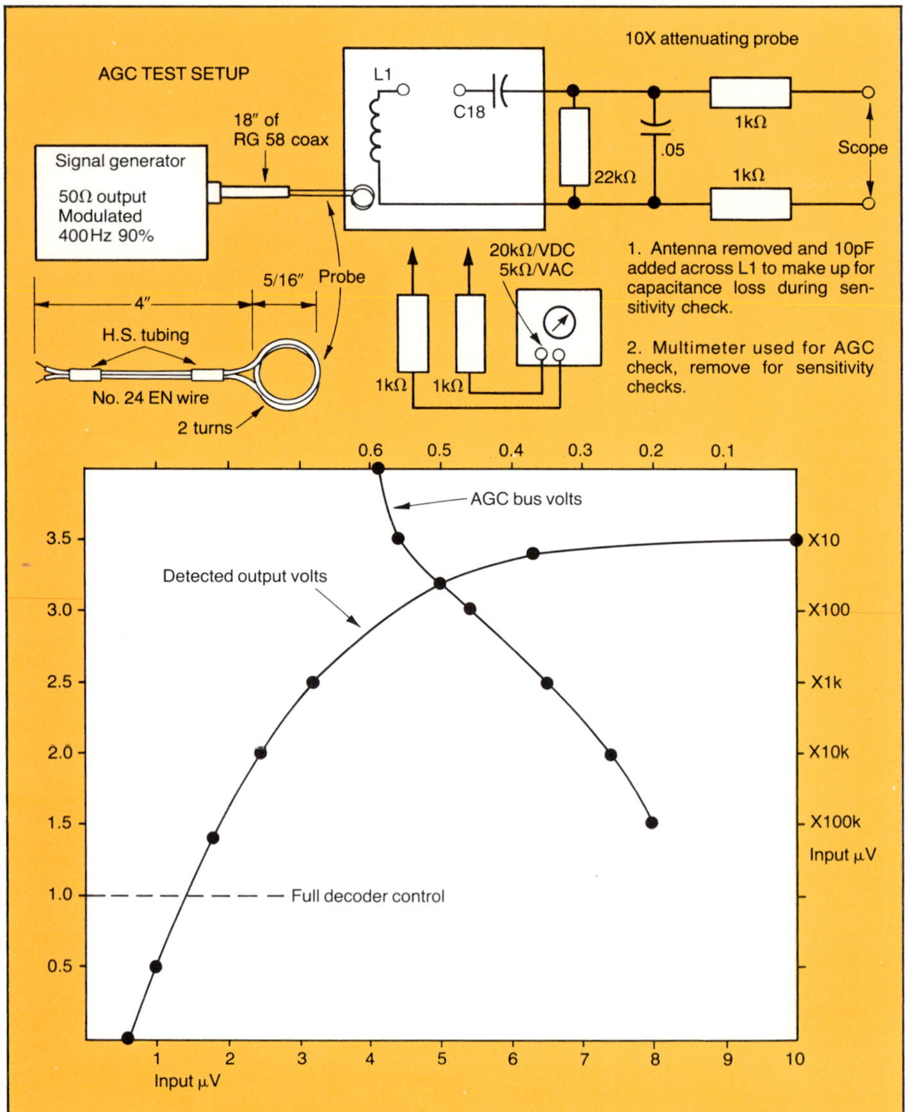

Fig. 3-2 AUTOMATIC GAIN CONTROL RESPONSE

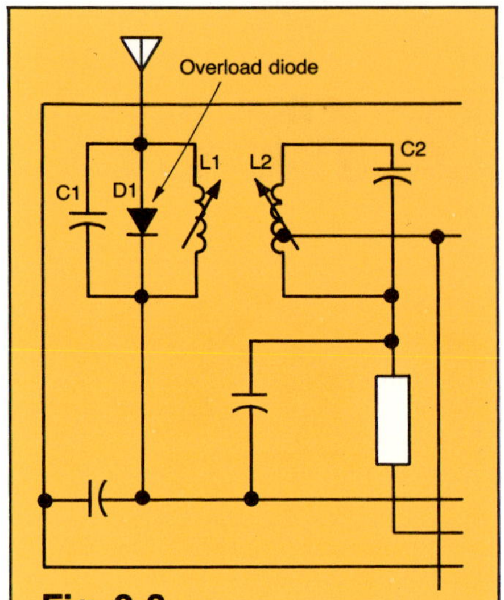

Fig. 3-3 RF OVERLOAD PROTECTION

used as the intermediate frequency in practically all RC superhets? It's a matter of economics — millions of tiny IF transformers on this frequency are used in pocket-size broadcast band receivers, so they are available from many manufacturers at low cost. Actually, a higher intermediate frequency would be a real advantage because it would move the troublesome image frequencies much farther from the RC signal frequencies. Even a single-tuned input circuit could then reject most signals on the image spots. But such sets would require specially made and much more expensive IF transformers, so manufacturers stick to 455 kHz. Higher intermediate frequencies would have a slight disadvantage in that the selectivity and amplification of each IF stage would be slightly less than at 455 kHz. However, an added IF stage could make up for this.

Most receivers use separate transistors for the local oscillator and the mixer. If these two functions are combined in a single transistor, the resulting mixer-oscillator is called an autodyne stage. It uses tuned circuits and a crystal to hold the receiver to the exact desired frequency, as does a receiver with separate mixer and oscillator transistors.

In either case, the mixer has an output of 455 kHz and this output is boosted in power by the intermediate-frequency amplifier (IF AMP). In most superhets, the IF AMP consists of three transformers, T1, T2, and T3, and two transistors. The latter are represented in Fig. 3-1 by the boxes labeled IF-1 and IF-2. Most of the amplification of the signal in a superhet occurs in the IF AMP, and this circuit also contributes most of the receiver's selectivity.

The detector is a diode or a transistor. It demodulates the modulated intermediate frequency from T3 (remember that at this point the IF is 455 kHz) into the audio frequency. You can hear the audio frequency at this point with a pair of sensitive headphones. You cannot, of course, hear a signal at any point in the circuit to the left of the detector. The detector also produces the automatic gain control voltage. This AGC voltage is fed back to IF-1 and IF-2, and often to the RF transistor, if any. It reduces receiver sensitivity when the incoming signal is strong so that a fairly even level of signal is always applied to the audio frequency amplifier. Figure 3-2 shows automatic gain control response curves and also shows the test setup used to measure AGC response. Some superhets have an additional circuit to reduce very strong signals — such as when you hold the transmitter antenna right next to the model. This overload circuit consists of a diode placed across the RF coils, Fig. 3-3.

Variations in receiver design have generally been differences in hardware rather than in concept. Several designs have used small devices called crystal filters that resonate at 455 kHz to improve receiver selectivity. Some have used three tuning coils in the front end, others have used a fourth IF transformer.

Decoders

Decoders have gone through essentially the same evolutionary pattern as encoders. The decoder performs as in Fig. 3-4. The receiver's output to the decoder must exceed the threshold level for the decoder input. Indeed, performance of a receiver is often specified as in Fig. 3-2 as "X microvolts for full control of the decoder." This means that a received signal of *X* microvolts will produce a detected signal that allows the decoder to operate reliably. The threshold level must be carefully chosen. Designers have used simple transistor switches, Schmitt triggers, one-shot multis, and, most recently, opamps to condition the decoder input. The input pulses must be square, clean, and must not vary in amplitude.

Further shaping and squaring may

be done by a clock formation circuit in the decoder whose output is precisely squared and shaped. The decoder's sync pulse network is an integrator whose time constant is greater than the maximum control pulse length of 2 ms but shorter than the sync pause of about 5 ms. Such a waveform usually appears at some intermediate stage in the decoder as a sawtooth pattern. A sophisticated decoder squares up the sawtooth. The decoder's reset, or clear, network has a long time constant so that it cannot go low as long as any part of a given frame appears. This is the same as in the encoder. Some early decoders used the reset circuit to trigger a fail-safe circuit to automatically center all servos, but this was abandoned when systems became more reliable. In logic terminology the decode function is termed series-in, parallel-out logic in that a serial bit stream is sequentially decoded as one of N where N is the last channel.

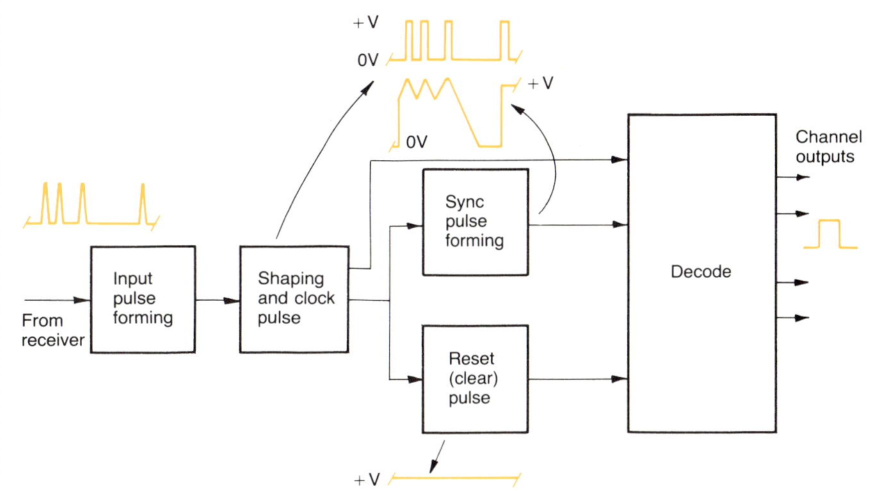

Fig. 3-4 DECODER BLOCK DIAGRAM

More complete understanding of the decoder comes from a study of the chronology of its development. Figure 3-5 presents a decoder from the mid-60's. Q7 through Q10 and related components are the pulse-forming networks. Q7 forms and shapes the receiver output pulses to form the decoder's clock, or shift, pulses, one for each detected pulse. Q8 inverts the pulse and Q9 and C90 integrate it, stretching it into a sawtooth. Q10 then squares up the set pulse.

Keep in mind that the first block, unit, flip-flop, or whatever, that decodes the first channel cannot switch states unless the sync pulse is low (or high, depending on the design). Thus the need for the sync pause. The same is true for all series-in, parallel-out devices. As the first clock, or shift, pulse enters the decoder, the set pulse is low for a few microseconds until the sync, or set, pulse begins to rise. Q12 and Q13 then switch states. (That decoder had a negative-going pulse so they switched low.) They cannot switch states again until the second shift pulse is received to complete the channel one control pulse. When the second pulse arrives, Q12 and Q13 revert to high state, but the sharp rise is coupled to Q14 and Q15 to trigger that pair to low state. Note that the pair Q12 and Q13, and other pairs of transistors, are connected in a configuration that is similar to a device known as a silicon controlled rectifier (SCR). The junction at the collector of Q12 and the base of Q13 is the gate. Once the switch is gated, it can only be turned off by disconnecting the emitter of Q12 from plus (the shift line sits high except during an incoming pulse). That is why the second pulse turns off the pair Q12 and Q13.

The shift pulse appears on the shift line each time a pulse is received. Why doesn't the pair Q12 and Q13 just sit there and cycle? That's where the sync, or set, pulse comes into play. As I mentioned, the sync pulse goes high just microseconds after the first shift pulse arrives. It remains high until well after the last shift pulse, and Q12 and Q13 can only shift state when the sync pulse is low. The sequence ripples down the line of switches just as it went from the first pair to the second. During the sync pause, the sync pulse circuit has time to switch states to set the pair Q12 and Q13 for the first pulse of a new frame. And so it goes at 50 frames per second.

SCR and SCS decoders

Many decoders have used the circuit just described. Others have used an SCR that performed the function of the two transistors in each pair. Bill Cannon chose such a design in the late 1960's. Here's Bill's description of that decoder (Fig. 3-6): "The first incoming pulse to the base of Q8 turns that stage off, and applies the resultant negative pulse through capacitor C23 to the gate circuit of SCR-1. Time constants of C22 and R24 in the base circuit of Q8 continue to hold this stage in the off condition until after the pulse train is completed. C22 resumes its normal charged state during the sync pause, turning on Q8 again. The decoder is then ready to process the next frame.

"Capacitor C21 and resistor R23 comprise a pulse-forming network that differentiates the incoming signal pulses and produces a series of short positive trigger pulses. Applied through the turn-on line and isolation diodes, these pulses turn on the SCRs. Note that pulse No. 1 of the pulse train simultaneously applies both a positive and a negative pulse to the gate of SCR-1, which is normally conducting. The longer duration negative pulse at the gate turns off SCR-1. Signal pulse No. 2 has no effect on Q8, which is now off, but applies a positive pulse to the gate of SCR-1 through diode CR-5, turning SCR-1 back on and producing the channel one output pulse. Conduction of SCR-1 causes a negative pulse to be applied through C24 and R27 to the gate of SCR-2, shutting it off. SCR-2 turns on when the gate is triggered by the positive pulse from the turn-on line caused by signal pulse No. 3. Thus, the channel 2 output pulse is produced. In the same way, all SCRs sequentially trigger off and on to obtain the desired number of output pulses (servo input signals)."

Ed Thompson pioneered a more versatile device, the silicon controlled switch (SCS), that performs the same function as the SCR but has greater flexibility because the SCS can be completely controlled at the gate. The flow of current need not be interrupted as with the SCR, resulting in simpler circuitry. Numerous sets used an SCS decoder, notably Heathkit and Royal. SCS devices were tricky because subtle changes in manufacturer specifications for the same device number could render them unsuitable at the 4.8-volt level used in receivers.

IC decoders

A few decoders used integrated circuits containing two J-K (a class of logic) flip-flops for decoding. However, these ICs were available with only two flip-flops per 14-pin IC, which is wasteful of PC board space, so they were soon supplanted by the plentiful, popular, eight-bit shift register that can decode up to eight channels on a single 14-pin IC. The eight-bit shift register can decode eight pieces (bits) of information. Its logic requires precisely the same clock, sync, and reset pulses used in earlier decoders, so it was a natural. Almost all decoders since 1973 except those in inexpensive two-channel sets

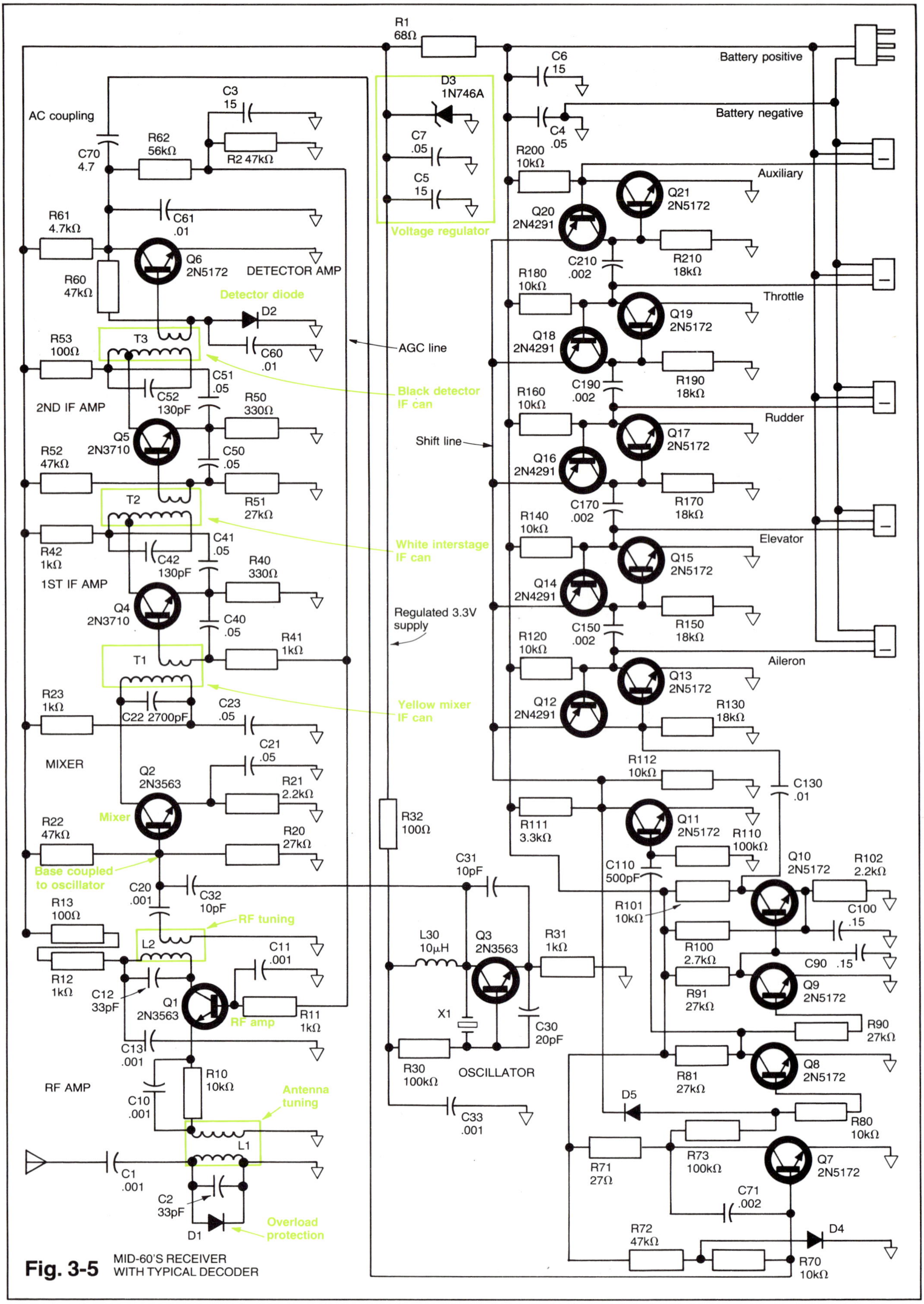

Fig. 3-5 MID-60'S RECEIVER WITH TYPICAL DECODER

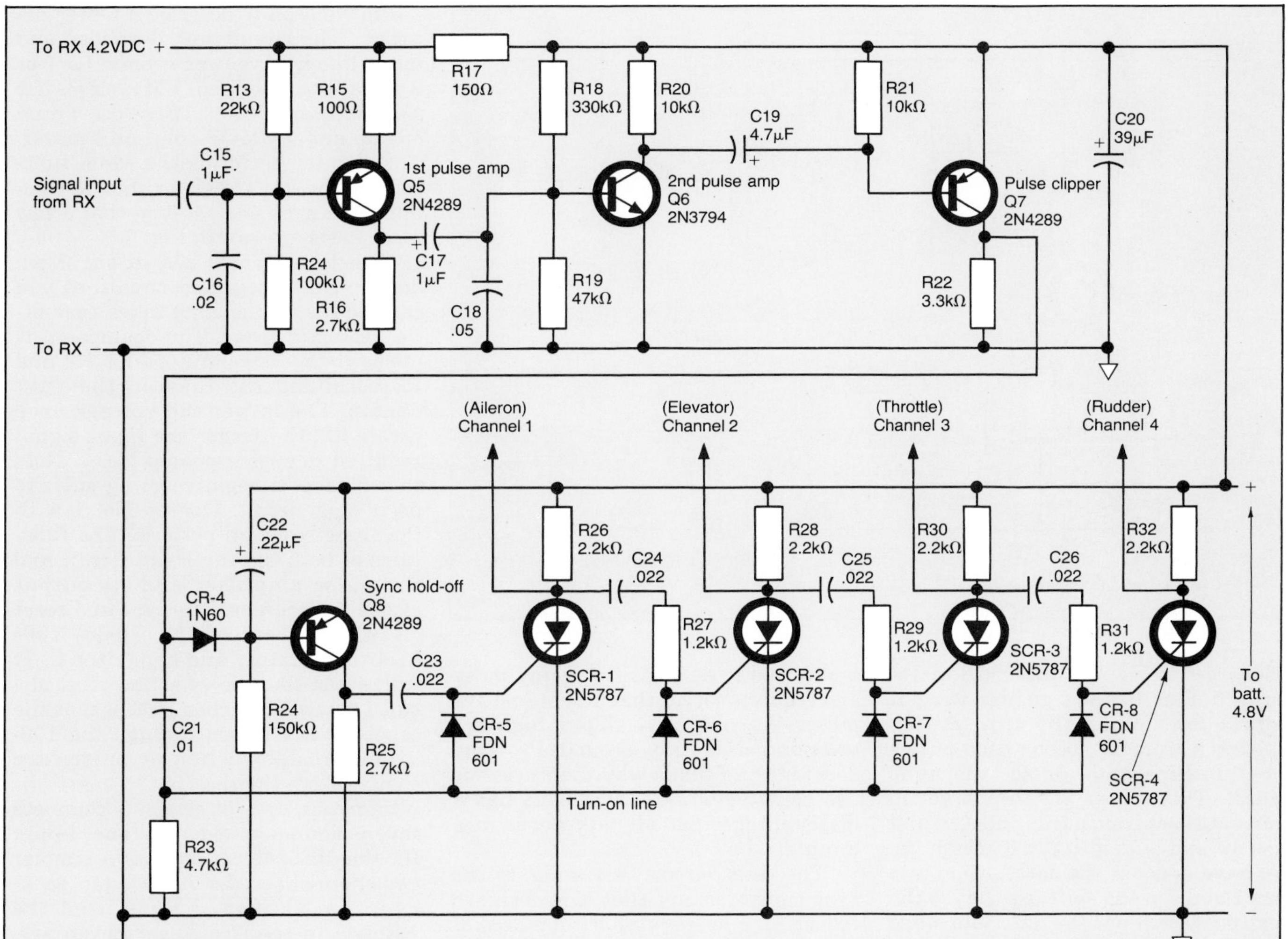

Fig. 3-6 CANNON DECODER

have used the SN74164 shift register or a similar IC. As technology has advanced, we've gone from TTL logic to CMOS logic that drains less than 4 mA and provides better noise rejection.

Some designers who adopted the shift register continued to use discrete components for the clock pulse, sync pulse, and reset circuits, while others soon began to use a hex inverter to do those jobs. A hex inverter contains six inverter transistors in one IC. I used the hex inverter and shift register in the Ace decoder in Fig. 3-7. Jim Oddino described a similar circuit (Fig. 3-8): "The transistor provides the threshold current. It is biased to the point that it is almost turned on by the diode and two resistors in the base circuit. Current flows through the diode and the 10-kilohm resistor which causes a 0.6-volt drop across the diode. We know that we need 0.6 volt across the base-emitter junction of a transistor to turn it on, so if we connect the diode right across the transistor, it will turn on. But we want it almost on, so we put a resistor between the diode and the base of the transistor. The transistor will now turn on when very small signals are applied to its base by the receiver detector. You can control how small by changing the 27-kilohm resistor. The larger the resistance, the larger the signal required to turn on the transistor.

"The next part may look complicated, but those six little triangles represent inverters which are all contained in one integrated circuit. The signal goes in on pin 1 and comes out on pin 12, slightly filtered by capacitor C1. It is basically the same shape as before but it is squared up and has constant amplitude as a result of being amplified in the two inverters.

"The sync circuit is made up of a diode, capacitor, and inverters 5 and 3. The input to this circuit must have the opposite polarity from the clock (it must be high when no pulses are coming from the receiver) so inverter 2 is inserted. As long as pulses are coming from the receiver, capacitor C2 will be

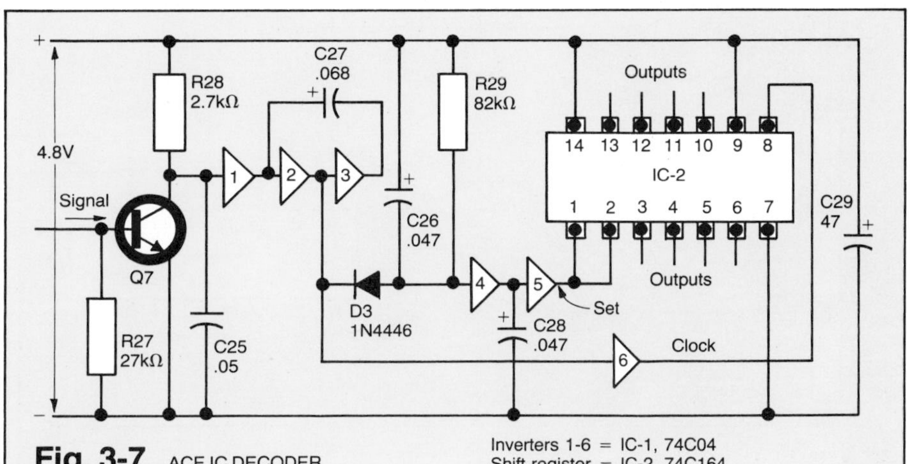

Fig. 3-7 ACE IC DECODER

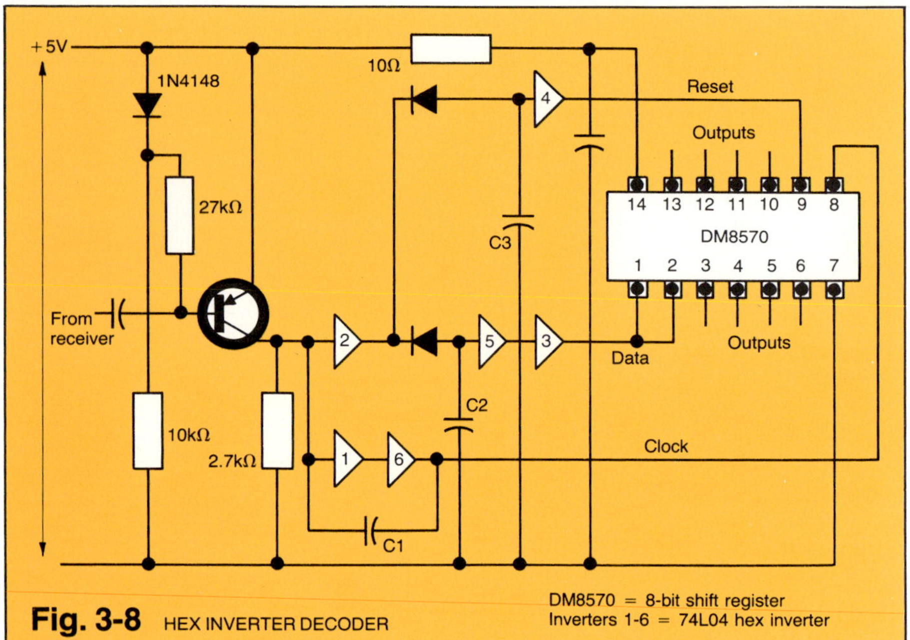

Fig. 3-8 HEX INVERTER DECODER

discharged below the threshold of inverter 5. Pins 10 and 5 go high and pin 6 goes low. When the transmitter finishes a frame it doesn't send pulses for 7 milliseconds or so [the sync pause]. This allows C2 to charge up (current flows from pin 11 into C2) and pins 10 and 5 go low, pin 6 goes high. We have now put the data, a one, into pins 1 and 2 of the shift register, so the next pulse will put the data into stage one. This comes out on pin 3 and is channel one. The next pulse shifts the data to stage 2 at pin 4 which is channel two, and so on. Note that as long as the pulses are 1 to 2 milliseconds apart, pin 6 is low and no new data can be entered. When I talk about data, I mean either a one or a zero represented by 5 volts or zero volt respectively. So when I say that the data is shifted, I mean pin 3 goes from zero to 5 volts and then from 5 volts to zero volt when the data is shifted to channel two. This is referred to as a positive-going pulse and requires servos that accept positive pulses. We could have all pins high at 5 volts and shift zeros around. Pro-Line does this and that's why their systems use negative-pulse servos. This has a disadvantage that I'll talk about in a minute.

"The reset circuit is similar to the sync circuit, except that it is designed to look for a longer absence of pulses. If no pulses come in, C3 charges up, pin 9 goes high, and pin 8 goes low, which causes all outputs to go to zero. This means your servos will not drive hard over if you turn off your transmitter or lose pulses for any other reason. Note that you can't use this feature if you have a negative-pulse system. If you reset all outputs to zero, all servos would drive hard over. You can get around this by AC coupling, that is, by placing a capacitor in the input to each servo."

Jim went on to describe a newer decoder: "The circuit just described and many like it have been around for 7 or 8 years. The new Kraft FM systems use a different approach. These use a dual opamp and a CMOS Johnson counter. The counter performs the same function as the shift register, but it combines the sync and reset so that a few more parts are saved, Fig. 3-9.

"The two opamps are in one 8-pin dual in-line integrated circuit, so you can see the whole thing takes very little PC board space. The opamps operate as voltage comparators. R1, R2, and R3 establish the bias on the first opamp. The larger the voltage drop across R2, the larger the input signal required to switch opamp No. 1. Note that it takes a negative-going pulse to drive this circuit. Opamp No. 1 with the three resistors performs the functions of both the threshold circuit and the pulse amplifier and its output clocks the counter. The sync and reset circuit consists of the diode, 100-kilohm resistor, and capacitor C. It works just like the sync and reset circuit I described earlier, except that the capacitor is charged through the 100-kilohm resistor when no pulses are coming from the receiver."

Signetics now produces a complete seven-channel decoder in one 14-pin IC, the NE5045, as well as a simpler two-channel decoder in an 8-pin package, the NE5046. I have used the NE5045 in the Ace Silver Seven system. Figure 3-10 presents a block diagram of the NE5045 decoder. The external components set the timing for the sync pulse, filter the regulator output, and set the minimum pulse width the decoder will accept. Control of the minimum pulse width is an important feature because the decoder can be set to reject noise spikes which are characteristically only a few microseconds long, compared to the 250-microsecond control pulse. This adds noise immunity. The sync pulse time is set to match the transmitter's frame time. The NE5045 is described in more detail in the next chapter.

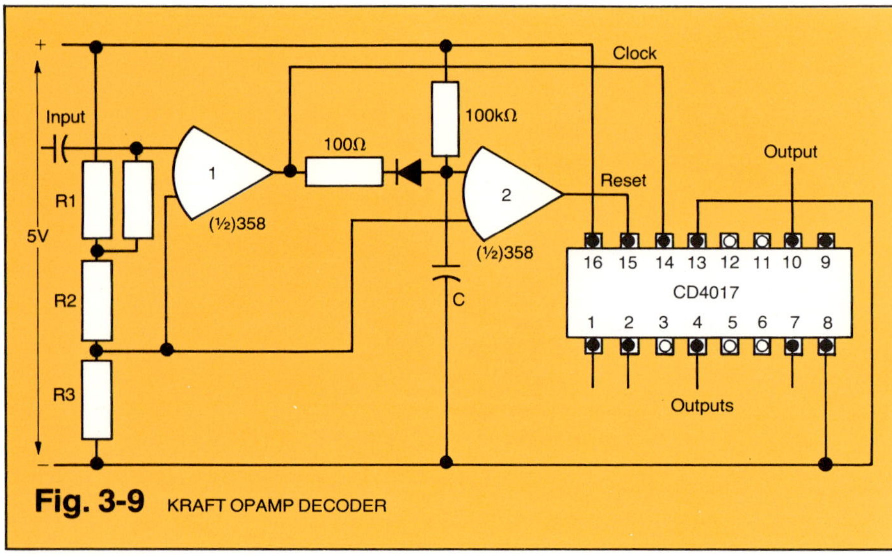

Fig. 3-9 KRAFT OPAMP DECODER

Servos

Servos provide the muscle for our systems. The mechanical parts of the servo consist of the case, gear train, output wheels and arms, feedback potentiometer element holder, and feedback potentiometer wiper holder. Although a few early servos had metal cases, 99 percent of the two to three million servos manufactured have been made entirely of injection-molded plastic except for a few metal parts such as gear retention pins, feedback wiper carrier, and screws.

The principal change in mechanical parts has been size, and size has been determined primarily by the diameters of small DC motors on the market. The earliest digital servos, affectionately known as "boxcars," were carry-overs from analog sets and used a 20 mm motor. Most servos over the years have used 16 mm motors. The Dunham Micro servo was the first to use a 12 mm motor, as do the D & R Bantam Midget servos. The Kraft KPS-18 and a recent Dunham miniature servo use motors approximately 10 mm in diameter. Not surprisingly, given the current trend to mammoth-scale aircraft, there's a new demand for more powerful servos, and we're back to 20 mm motors.

The servo gear train converts the motor output torque from a very low level to the highest level of torque needed to move the control surfaces. Gear trains have changed little over the years, although better plastics and improved injection-molding techniques have increased reliability and life expectancy.

In almost all servos the final output gear and shaft are attached to the feedback control wiper. In this book, I'll mean a potentiometer when referring to feedback. The first Heathkit servo, which was based on the Kraft KPS-9 mechanism, had a capacitor for feedback, and F&M developed a magnetic feedback arrangement. Neither of these feedback arrangements is used today.

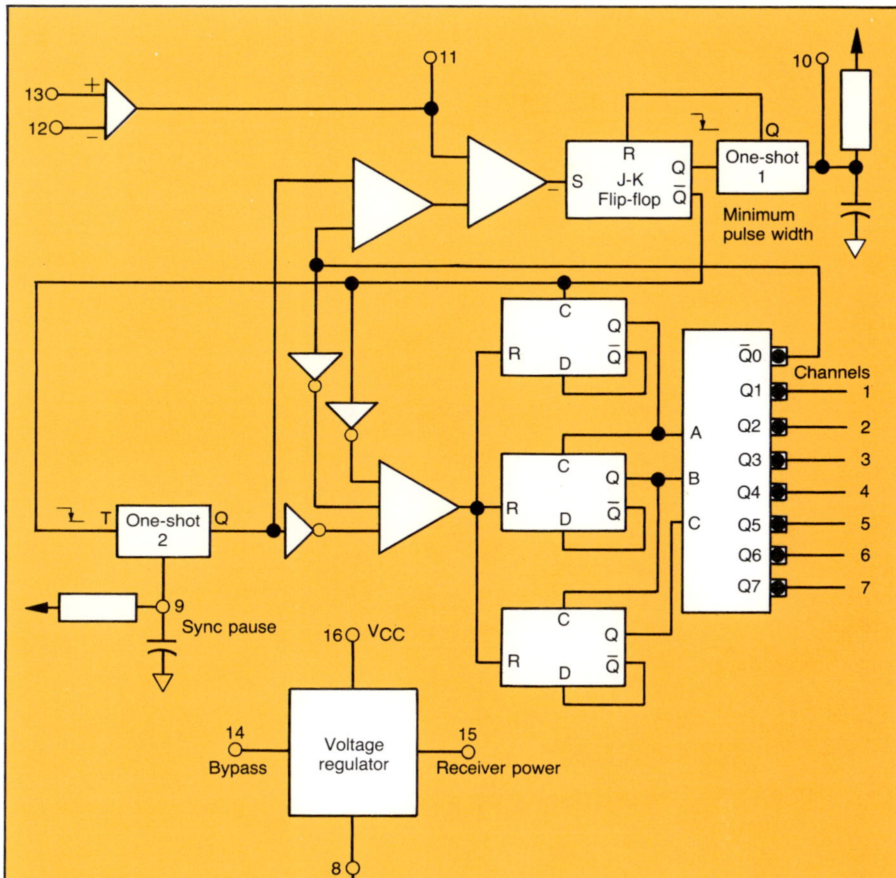

Fig. 3-10 NE5045 DECODER BLOCK DIAGRAM

Servo electronics

Review Fig. 2-7 and note that the servo amplifier must accept and respond to a square pulse that is 1.5 ms ± 0.5 ms long. At least 90 percent of all contemporary systems generate a positive-going pulse at the decoder outputs. Pro-Line, some EK systems, and early World Engines systems had a negative-going control pulse. To repeat Jim Oddino's definition of negative-going and positive-going pulses in slightly different terms, by a positive-going pulse, we mean that when viewed on an oscilloscope, the servo input is low (near zero volt) for all of the approximately 20 ms long frame except when its control pulse arrives. The input then rises abruptly to near plus 4.8 volts for the duration of the control pulse.

Figure 3-11 presents a block diagram of the servo function. I have used the positive-going input pulse for illustration. Figure 3-12 is a 1970 version of the Orbit discrete component servo. It is typical of the period 1965 to 1975. The servo accepts the control pulse that is 1.5 ms ± 0.5 ms long, generates a reference pulse for comparison that is based on the position of the gear-train-driven feedback pot, measures the difference in the length of the two pulses, and generates drive pulses to the motor to drive the gear train and the feedback pot. When the difference in pulse length is nulled, that is, it becomes zero, the servo stops moving. Subtleties of design involve the speed with which this can be made to happen, the amount of overshoot beyond null that the servo exhibits, and the output torque.

Refer to Figs. 3-11 and 3-12. The pulse arrives and passes trough R10 to the junction of R9. At the same time, the leading edge of the pulse is differentiated by C2 and R6 to form a trigger spike. Q1, Q2, and related components form a one-shot multi. The trigger spike causes the multi to switch states so that the voltage at the collector of Q2 goes low. The position of feedback pot R1 determines the time that the output of Q2 is low. As long as the control input isn't being changed by stick movement, the input pulse and the reference pulse will be equal within a microsecond or two. The difference in these times is called dead band and must be set close enough for tight servo positioning, but not so close that the servo buzzes.

R9 and R10 serve as a simple comparator. The waveforms in Fig. 3-13 represent a positive-going error, that is, the input pulse is longer than the reference pulse by 5 microseconds. Remember that the entire pulse length is

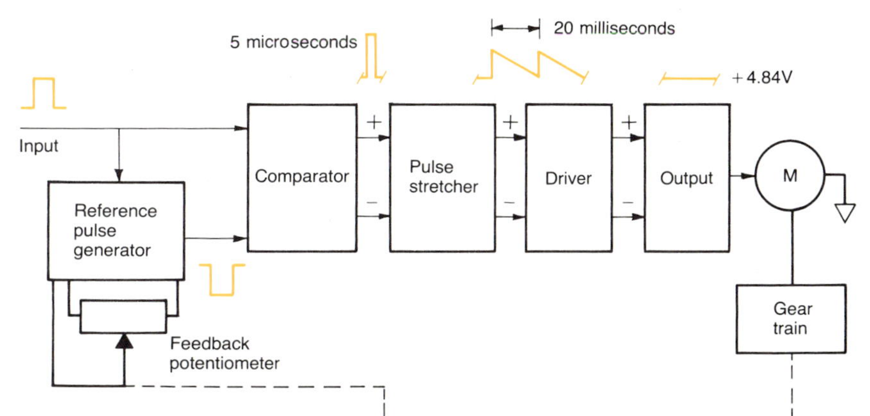

Fig. 3-11 SERVO FUNCTION

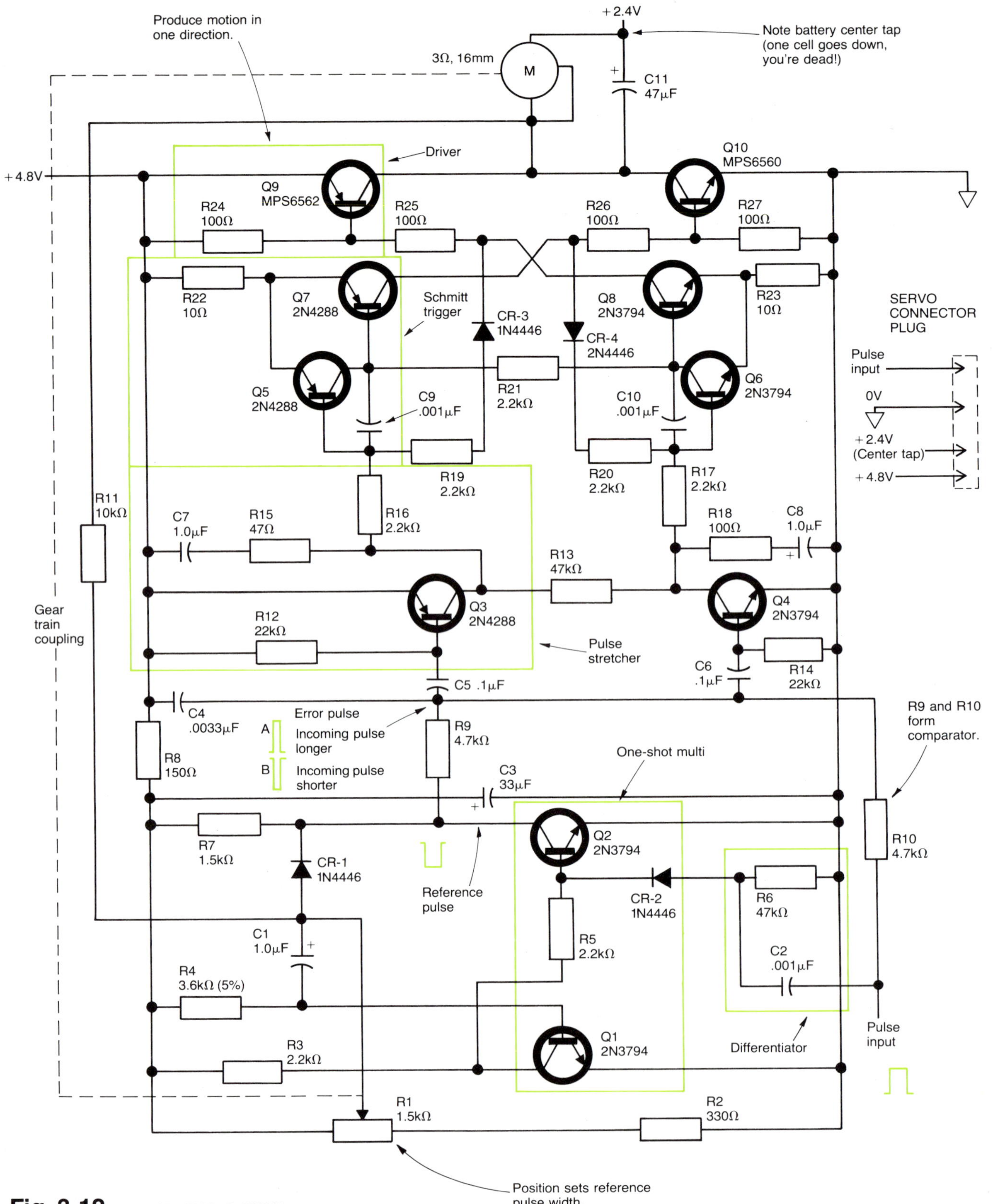

Fig. 3-12 1970 ORBIT PS4-D SERVO

1500 microseconds, or 1.5 ms, and that this sequence of input occurs approximately once each 20 ms. Three to five microseconds is the equivalent of 5 in 1000 (± 0.5 ms control motion) or 0.5 percent. Translated to a total stick motion of ± 30 degrees (60 degrees total), that means the servo will respond to 0.3 degree of stick motion. These figures are typical of contemporary servos — most older servos would have ignored our 5-microsecond error pulse. New IC servo amps respond to an error pulse of less than 2 microseconds.

If you were to send the 5-microsecond error pulse through the motor, nothing would happen except that the motor might hum slightly. We must stretch and amplify the input error pulse to generate a meaningful drive pulse. The pulse stretcher is an integrator. Our 5-microsecond error pulse is coupled to the base of Q4 via C6. If the error pulse is negative (the reference pulse is longer than the incoming pulse), the signal passes via C5 down the upper train of transistors to drive the servo in an opposite direction than it is driven by the positive error signal. When Q3 or Q4 is triggered, the appropriate capacitor, C7 or C8, is fully charged to integrate the pulse. As the error signal increases or decreases, the stretcher capacitor's discharge time occupies proportionately more or less of the 20 ms frame. The driver stage senses the amplitude of the stretcher ramp and produces a supply voltage amplitude pulse that reflects the stretched error pulse. That pulse may be a full 20 ms if there is sufficient error pulse duration. The driver drives the output transistor, Q9 or Q10, to full on, placing full available voltage across the motor for as long as there is an error pulse.

Note that I just stated "available" voltage. That is a principal point of distinction for servo design history. The supply of voltage for some of the earliest sets, for example, Bonner, Logictrol, and Digicon, was strange by today's standards. These sets used germanium transistors for the output because germaniums have low current drop and high current handling capability at saturation (full on). However, germaniums require a greater negative bias than silicon transistors for temperature stability. Thus, six nickel-cadmium cells totaling 7.2 volts at 1.2 volts per cell were used with the hookup in Fig. 3-14. The actual motor voltage was ± 2.4 volts measured from the ground side of the motor. Since the mid-1960's, all sets have used a 4.8-volt supply for all functions in the receiver and servos as in Fig. 3-12. However, the Orbit servo in that figure reflects the use of a four-wire, center-tapped 4.8-volt pack. Thus, the available supply voltage was a maximum 2.4 volts. The motors were usually of 3-ohms resistance.

The next breakthrough was the bridge output circuit developed by Bob Elliot of Logictrol for the MM-3 servo, which introduced the "three-wire" servo. In that design, a set of six transistors replaced Q9 and Q10 to place a full 4.8 volts across the motor and did away with the center tap, the fourth wire. Naturally, this reduced wiring and improved reliability. Few others adopted the bridge amp because it increases parts count significantly.

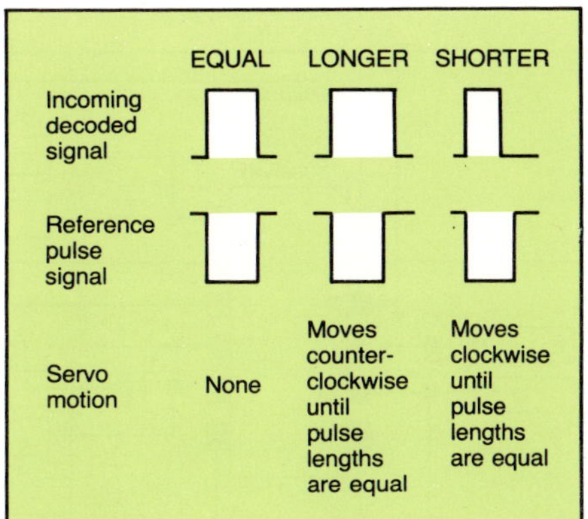

Fig. 3-13 SERVO INPUT RESPONSE

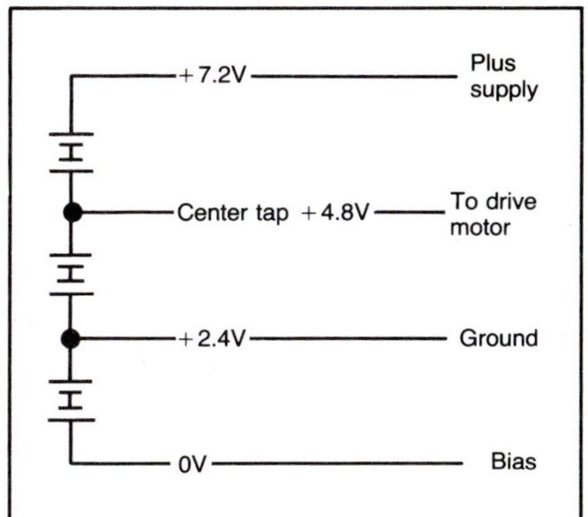

Fig. 3-14 EARLY SERVO POWER SUPPLY

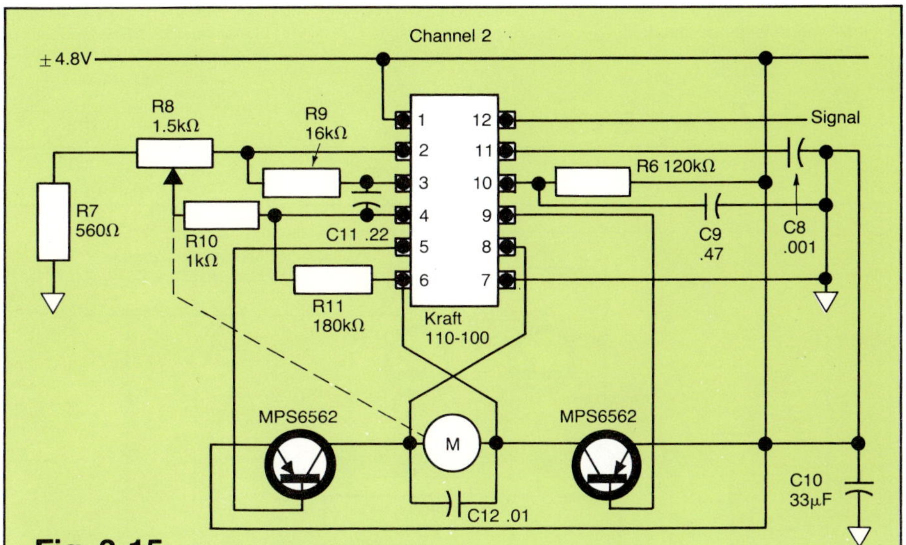

Fig. 3-15 KRAFT IC SERVO AMPLIFIER

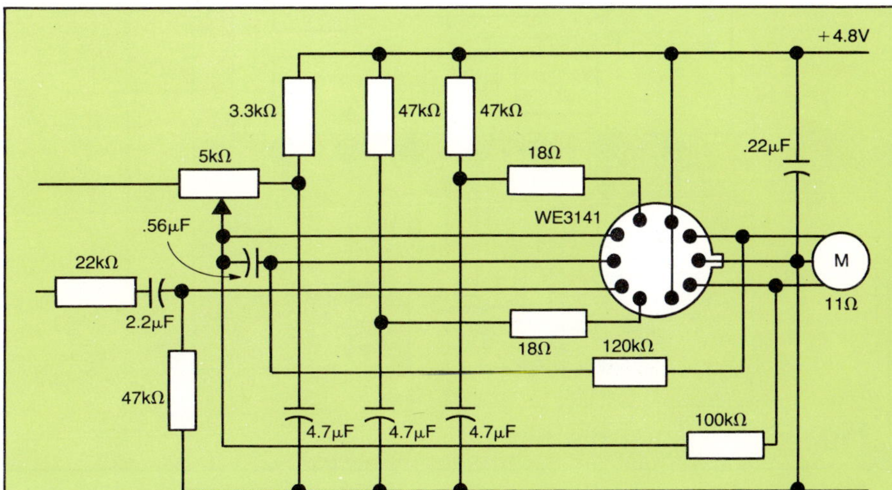

Fig. 3-16 NE543 SERVO AMPLIFIER

IC servo amplifiers

The next generation of servo amps saw the introduction of the special IC servo amp. The first was designed and produced by Signetics, the NE543. Texas Instruments produced a similar IC for Kraft, which was soon used with various modifications by several other manufacturers, Fig. 3-15. All servos with IC servo amps use 4.8 volts, have internal bridge amplifiers, and require a 5000-ohm feedback pot. Most accept only a positive-going pulse. Figure 3-16 shows a servo amp based on the NE543 that was used in the earliest Ace digital systems. Figure 3-17 presents a 1973 design based on the Texas Instruments SN28622. These amps use a motor of 8- to 12-ohms resistance. External drive transistors can be used with the SN28622.

In 1976, Signetics brought out a second generation servo amp IC, the NE544, that introduced the concept of the linear-ramp one-shot to servo de-

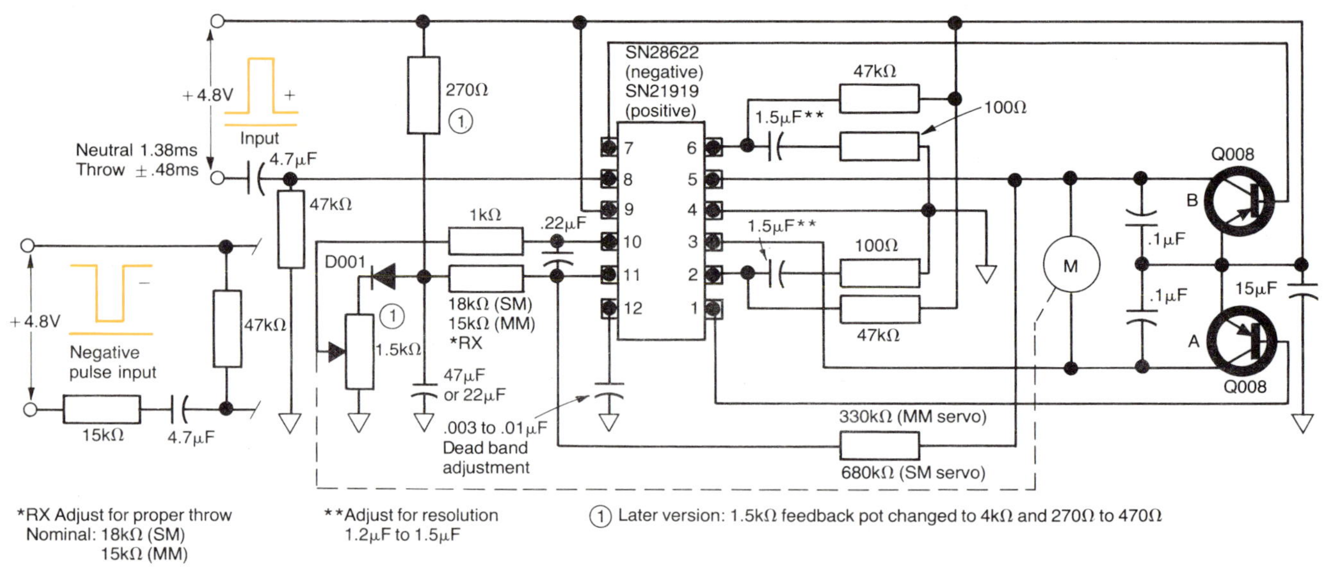

Fig. 3-17 SN28622 SERVO AMPLIFIER

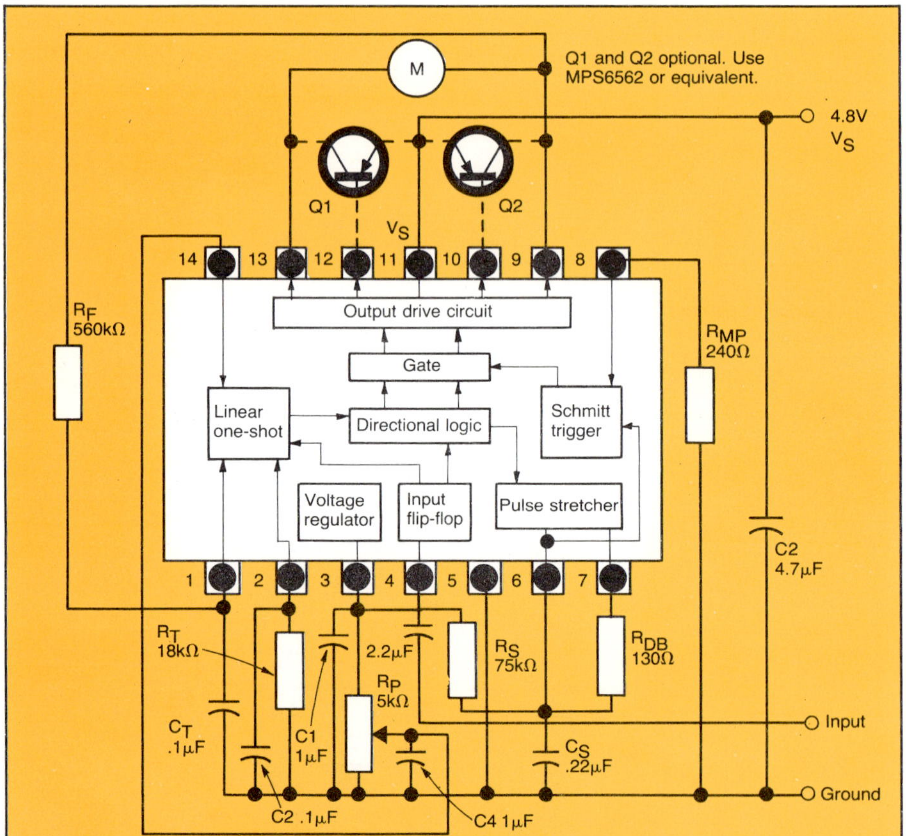

Fig. 3-18 NE544 BLOCK DIAGRAM

Positive Pulse Systems	Negative Pulse Systems
Ace Digital Commander	Controlaire
Blue Max	EK (Super Pro to 1973; white and black nonpolarized plugs)
EK (1973 and later; red and black polarized connectors)	EK-Logictrol II
Futaba	EK MM3 servos
Heath	F & M
Kraft	M. A. N.
Micro Avionics	OS
MRC	Pro-Line
Orbit	
Royal	
Sanwa	
World Engines Midget	

Table 3-2 POSITIVE AND NEGATIVE PULSE SYSTEMS

sign. To that time, as I pointed out in the discussion on encoders, servos depended on nonlinear charging of a capacitor to set the time constant in the reference generator. The NE544 incorporates a precision current source that yields linear charging. Figure 3-18 presents the block diagram for the NE544. Note that the functions are handled slightly differently than in the traditional amp since the IC permits the use of gates and logic to steer the error pulses. Ace first used the NE544 in 1977. All other RC manufacturers now use the NE544 or the Texas Instruments IC.

In the NE544 servo amp, a positive input signal applied to pin 4 sets the input flip-flop and starts the one-shot time period. The directional logic compares the length of the input pulse to that of the internal one-shot and stores the result of this comparison in a directional flip-flop. The exact difference in pulse width between input and internal one-shot pulse, the error pulse, is also fed to pulse stretcher, dead band, and trigger circuits. These circuits determine three important parameters:

- Dead band — The minimum difference between input pulse and internally generated pulse that will turn on the output.
- Minimum output pulse — The smallest output pulse that can be generated from the trigger circuit.
- Pulse-stretcher gain — The relationship between error pulse and output pulse. These parameters are adjusted by external resistors and capacitors at pins 6, 7, and 8. Dead band is controlled by resistor R_{DB}. Minimum output pulse is controlled by R_{MP}. Pulse-stretcher gain is adjusted by capacitor C_S and resistor R_S. The trigger circuit activates the gate for a precise

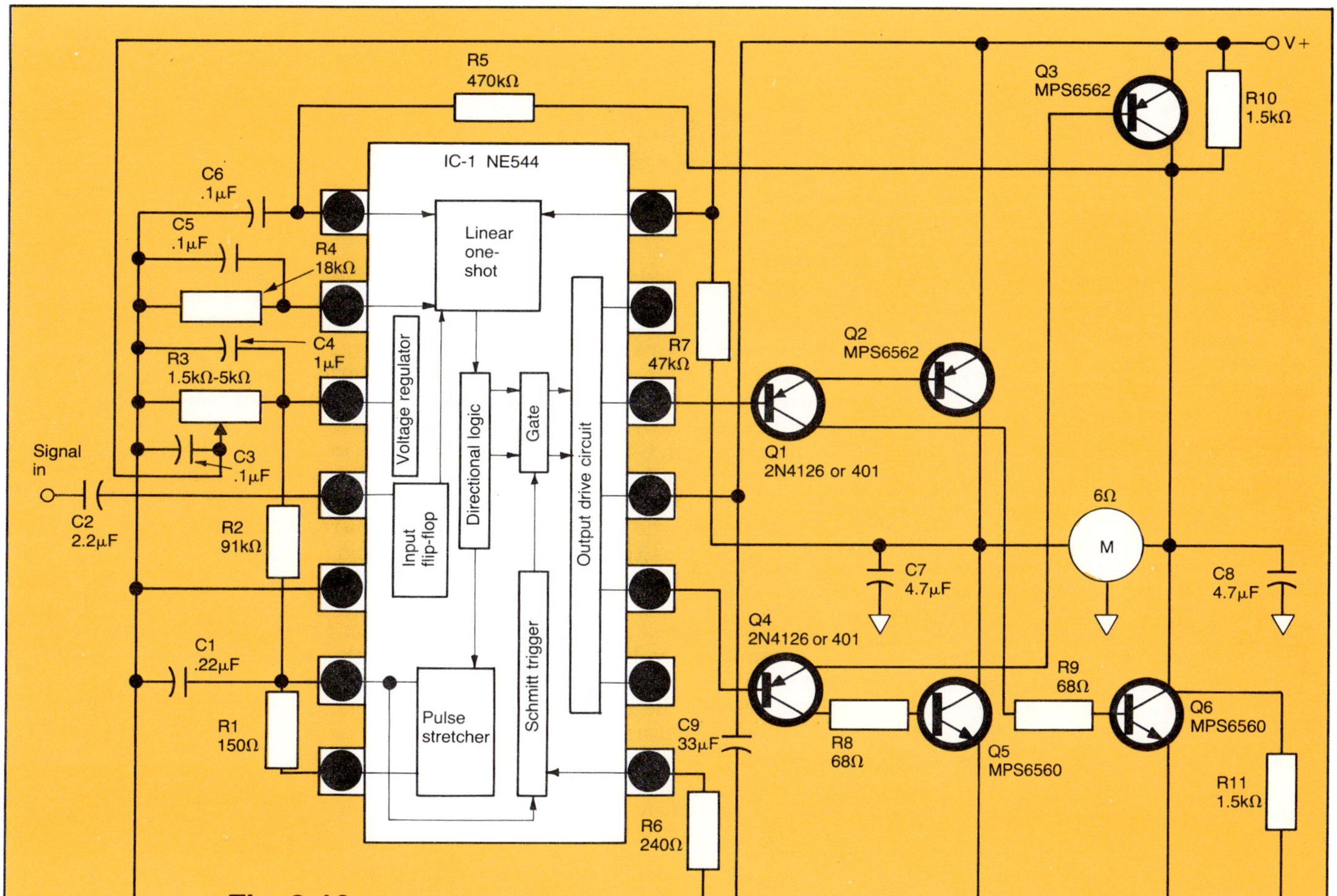

Fig. 3-19 EXTERNAL BRIDGE CIRCUIT

length of time to provide drive to the bridge output circuitry in proportion to the length of the error pulse. Resistor R_F determines the amount of feedback required for good closed-loop damping. Q1 and Q2 are external PNP transistors for increased motor drive, which makes for a faster, more powerful servo with better resolution. The amount of servo travel is controlled by resistor R_T.

The NE544 accepts only a positive input pulse. If you have a system that uses negative input pulses, but want to use servos with the NE544, you can buy a pulse inverter from Ace and several other manufacturers. Table 3-2 lists the pulse polarity of a number of systems, some of which are no longer manufactured.

The NE544's motor drive circuit permits the NE544 to be used as a complete servo amp. External drive transistors aren't required. This reduces expense because fewer parts are used. However, for high performance applications, the NE544 motor drive can be bypassed externally, Fig. 3-18, with a discrete PNP transistor. For even higher power output, such as that desired in 1/3-scale models, you can use a completely external high-power bridge circuit, Fig. 3-19.

That ends our trek through the RC system. There have been many contributors to the development of digital proportional RC systems for models, some of whom I've mentioned in passing. We modelers are indebted to Pullen, Spreng, Mathes, Dunham, Rehling, Cannon, Elliot, Fosgate, and others, for their pioneering efforts.

4 Super system

Let's look now at a contemporary super system to see even more clearly how integrated circuits using designs and concepts borrowed from computer engineering have revolutionized RC systems. The Ace Silver Seven that we'll examine is a seven-channel set so-named to celebrate Ace R/C's 25th anniversary.

Encoder

The Signetics NE5044 seven-channel encoder that was described in general terms in Chapter 2 makes possible most of the special features in the Silver Seven. A schematic for the simplest encoder using the NE5044 appears in Fig. 4-1. Values in Fig. 4-2 for the NE5044 are nominal values from the Signetics applications note. Those in Fig. 4-3 are the values I chose for the Silver Seven.

Here's how the IC is wired: Pins 1 through 7 of the NE5044 accept the seven control inputs. Pin 8 is ground (zero volt). At pin 9, R2 and C6 set the frame time at a nominal, fixed 20 ms. At pin 10, C5 and R3 set the output modulation pulse width at a nominal 250 microseconds. A 250-microsecond ramp appears at pin 10 and that pulse reappears as a full 5-V regulated amplitude square pulse at pin 11. At pin 11, R4 and R5 form a network that sets the bias for the base of Q1. The divider network formed by R6, R7, and R11 at pin 12 sets the maximum throw range at the desired $\pm$ 0.5 ms from a nominal 1.5 ms. The 100-kilohm resistor in series with pin 12 damps any RF. The series resistance set by R8 and R10 at pin 13 interacts with C4 at pin 14 to set the output control pulse width (not the modulation pulse) to a nominal 1.5 ms. C4 is identified on the Signetics block diagram, Fig. 4-1, as the C multiplex, C_{MUX}. C9 is part of the stabilization circuit for the constant-current source in the IC. Pin 15 presents a regulated output of a nominal 5 volts that we have labeled V_R. This is the plus supply voltage for the rest of the encoder. C2 and C3 are filter capacitors. Pin 16 is the 9.6-volt supply voltage. R9 ensures that the regulator starts properly.

The internal workings of the encoder are easier to follow if you study Fig. 4-1. The multiplexer functions as a strobed (clocked) voltage follower so that each input, when strobed, appears as a high-impedance input of more than 1 megohm and transfers the channel input voltage to the output in sequence. Only one of the seven inputs is active at any time. The high input impedance permits the flexibility of input functions described later.

The constant-current generator is a bidirectional current source whose current, I_C, is set at a constant, precisely controlled level regardless of the state of charge of C_{MUX}. The current generator alternately charges and discharges C_{MUX}. An internal feedback loop maintains the constant current and the high output impedance. This yields a typical linearity error of voltage input to pulse width output of less than 0.1 percent. C_I ensures stability of the feedback loop and in conjunction with R_I sets the current rate.

Two high-gain comparators, C1 and C2, compare the voltage across C_{MUX} with the multiplexer output voltage and the range input voltage. The comparators feed the counter control logic, which in turn controls the counter and current generator. The operation of this loop is as follows: When I_C is positive (sourced from the current generator into C_{MUX}) the capacitor charges linearly until it reaches a voltage equal to the multiplexer output voltage; assume this to be the voltage at pin 1, V1. At this time the output of C1 goes high, which reverses the direction of I_C (sinking into the current generator from C_{MUX}). C_{MUX} now discharges linearly until it reaches the voltage set on pin 12, V-range. At this time the output of C2 goes high, which again reverses the polarity of I_C, clocks the counter, and triggers the output one-shot. C_{MUX} again charges, but now C1 goes high when C_{MUX} reaches V2, the voltage on pin 2. The resulting voltage waveform is a triangle wave whose positive peaks correspond to the voltages on pins 1 through 7 for the first through seventh peaks and whose negative peaks are constant and equal to V-range. This waveform is in the first portion of Fig. 4-4. Independent control of I_C and V-range allows the encoder to be tailored to the desired combination of input voltage changes and output pulse width changes.

The frame generator, operating as a free-running multivibrator, controls the encoder frame time. The encoder generates a synchronizing pulse at the end of each frame. When C_{MUX} reaches the seventh positive peak, it reverses and discharges to V-range. The counter is clocked to the state where Q_O is high when VC_{MUX} equals V-range. C_{MUX} again charges, but now the output of C1 is ignored because Q_O is high, and charges to V-clamp and remains there. The encoder remains in this state until it receives a pulse from the frame gen-

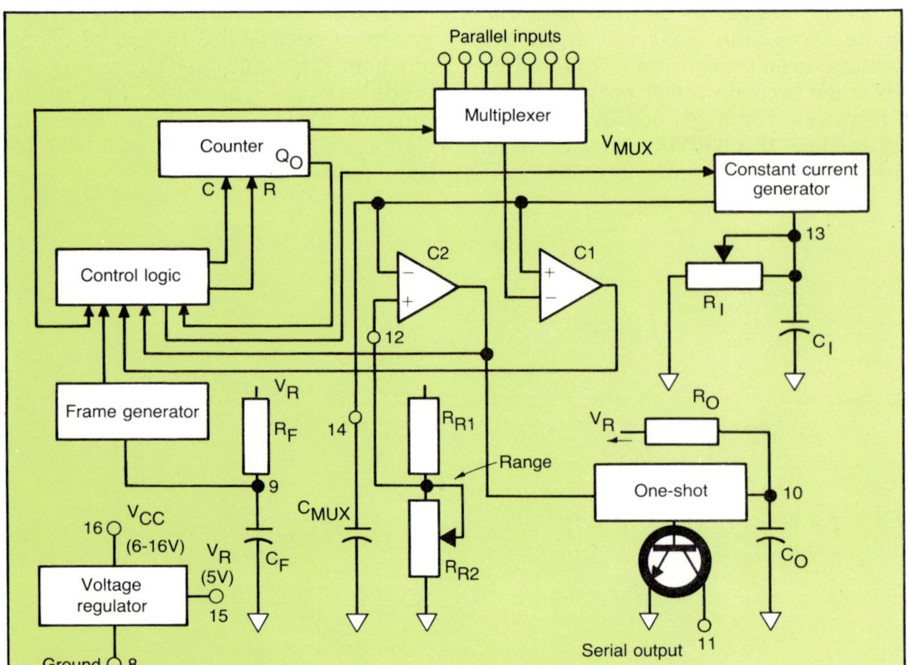

Fig. 4-1 NE5044 ENCODER BLOCK DIAGRAM

erator. With R_F and C_F connected, the frame generator operates in the astable mode, producing a narrow-pulse output. This pulse allows C_{MUX} to start discharging again. When C_{MUX} reaches V-range, the counter is clocked to the state where Q_O, channel 1, is high and the entire process starts over. The frame rate in this mode is .66 times R_FC_F and is referred to as the fixed-frame mode. I chose a frame time of 20 ms for the Silver Seven to match the optimum time constants of the NE544 servo amplifier IC. The output one-shot generates a positive pulse whose width is equal to R_O times C_O. The output is an open-collector NPN transistor capable of sinking 25 mA.

The Silver Seven encoder contains more components than the block diagram identifies. Here is a correlation of those values:

Block Diagram		Silver Seven Schematic
C_O	=	C5
C_I	=	C9
C_F	=	C6
C_{MUX}	=	C4

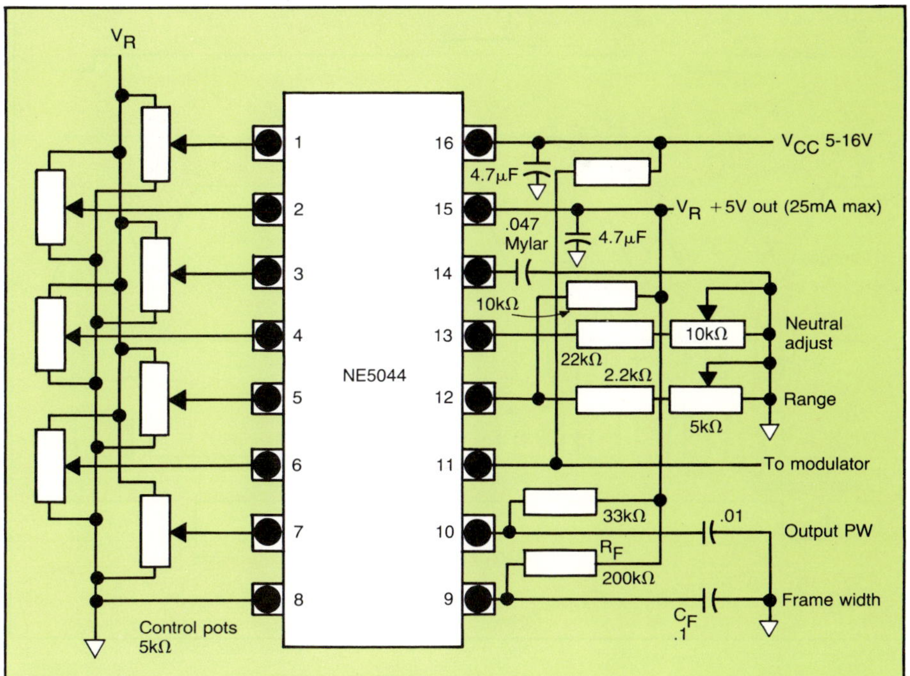

Fig. 4-2 NOMINAL VALUES FOR THE NE5044

Looking now at the left side of Fig. 4-3, C8 shapes the output pulse from pin 11 of the IC. Q1 inverts the pulse and amplifies it to about 3.5 volts peak-to-peak to modulate the RF section. The pulse is further sloped by C10 to make the modulation pulse conform to the desired shape.

The input to each channel may take any of many forms. The only requirement is that each channel's input be conditioned to have the same end voltage, nominally 2.5 V ± 0.5 V. Figure 4-5 illustrates the forms of input for the Silver Seven. The required 5-volt bus from pin 15 of the NE5044 is referred to as V_R in the figure and in the discussion below.

Buddy box

The modulation output pulse is also routed to the buddy box connection, Fig. 4-6. When the student end of the cable is connected to the student's transmitter:

- Pin 4 is jumpered to pin 3 via the student connector end of the cable.
- Pin 3 goes straight to the modulation output from the encoder (point B) and shorts RF modulation to ground, killing the student's RF.
- The student's encoder continues to operate with its output from point A separated from modulation out by the 2.7-kilohm resistor R13.
- The student's encoder is routed to pin 1 (wire 2) of the buddy box connector and then to the trainer's transmitter.

Looking now at the trainer's transmitter:

- Pin 4 is ground.
- The input from the student's encoder arrives at pin 2 where it connects to the 2.7-kilohm resistor R14 and is routed to the normally closed (NC) connection of the enable switch.

The instructor has control when the enable switch is at NC because:

- The student's encoder input goes straight to ground via common on the enable switch.
- Point A of the trainer's encoder board is open as far as the buddy box is concerned.
- Point B is connected to ground via R14 through the NC connection on the enable switch and the trainer's transmitter configuration is essentially the same as if there were no buddy box.

The student has control when the enable switch is depressed because:

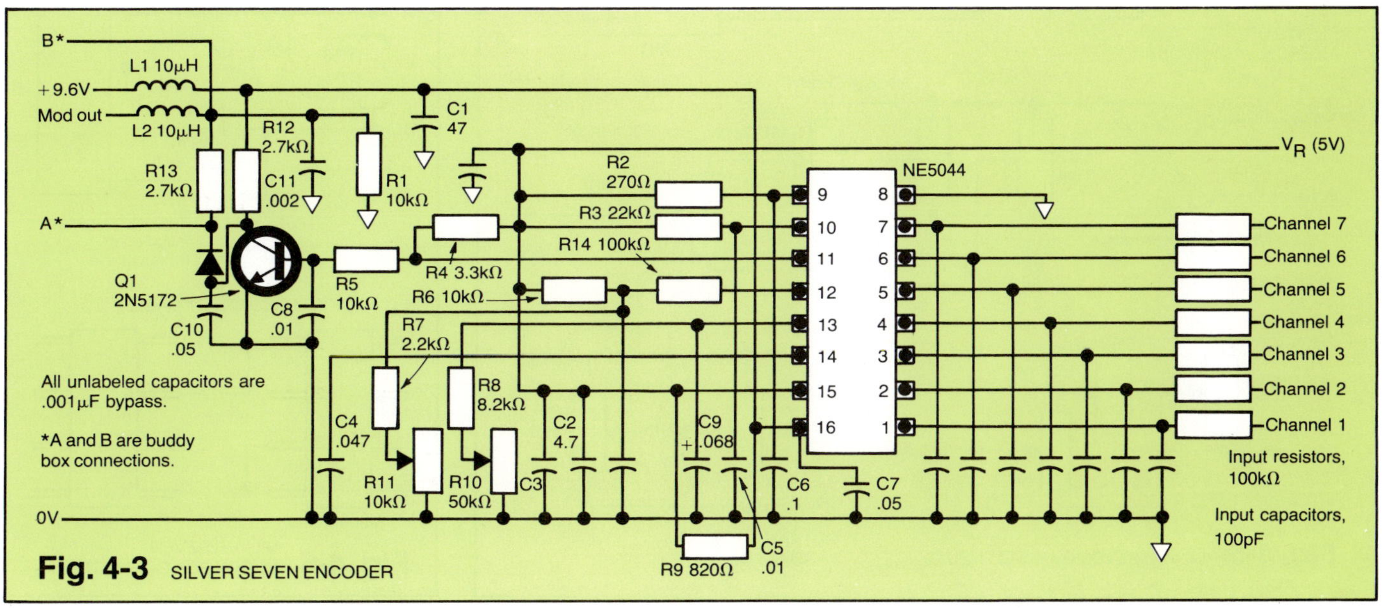

Fig. 4-3 SILVER SEVEN ENCODER

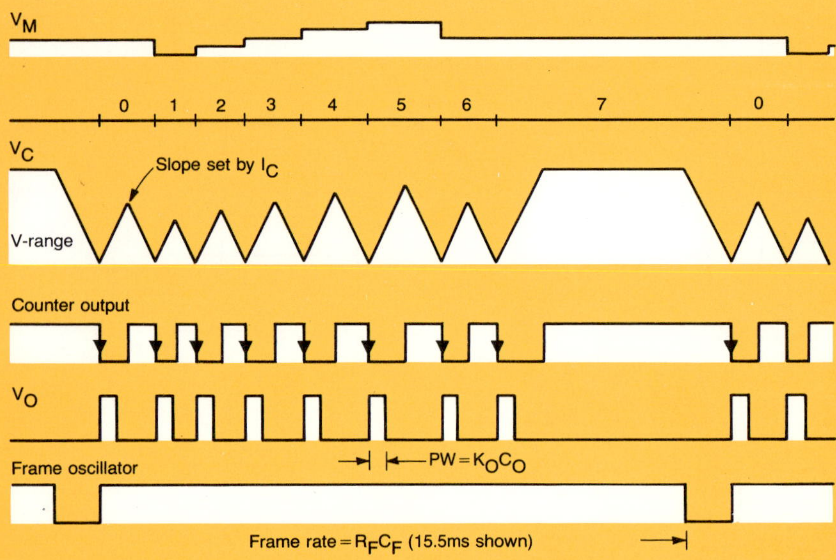

Fig. 4-4 SILVER SEVEN ENCODER TIMING DIAGRAM

- The output from the trainer's encoder (point A) is grounded via the normally open (NO) connection of the enable switch.
- At the same time, as the enable switch comes off the NC contact, the student's input is no longer grounded.
- The student's encoder pulse train is routed via R14 to point B (trainer's modulation output) to control the trainer's RF modulator.
- Point B is now grounded via R13 to point A and NO of the enable switch. (The student must never depress his enable switch.) Thus, the apparent configuration of the student's transmitter is essentially the same as if there were no buddy box.

Controls

The basic Silver Seven uses D&R control sticks so that control function and control trim are achieved by separate potentiometers, one connected to the stick and the other connected to a small tab at the edge of the stick. Precision metal open-gimbal sticks are an option. Each stick has an additional tab for an auxiliary control. In this way, each of the two sticks has three functions, accounting for six of the seven functions. The elements of all five potentiometers on each stick have a resistance of 5000 ohms. The seventh function, channel 5, is the retractable landing gear switch.

The basic functions provided by the circuitry on the transmitter PC board are control input; trim input; variable channel throw for full-rate; dual-rate and low-rate throw adjust on channels 1, 2, and 3; variable trim authority for channels 1, 2, and 3; and throw-reversal switches on channels 1, 2, 3, and 4. As I mentioned, channel 5 is the retract switch — each end is adjustable from 1.0 ms to 2.0 ms. Channels 6 and 7 are identical and have only center-adjust and throw-adjust. Center-adjust is important on channels 4, 6, and 7 because it allows for offset of the servos, say for flaps or throttle. A more suitable alternative is provided for channels 1, 2, and 3, and for convenient setup of the mixer and nonlinear modules which demand that all inputs be at precisely V_R divided by two.

The V_R/2-divider, Fig. 4-7, is an opamp connected as a voltage follower. The 22-kilohm resistors have a tolerance of 1 percent and divide V_R with that precision to form a reference for the plus input of the opamp. The output from the opamp is fed back to the minus input so that the opamp acts as a high-gain amplifier that keeps the output locked precisely to the plus reference input. Visualize that if V_R

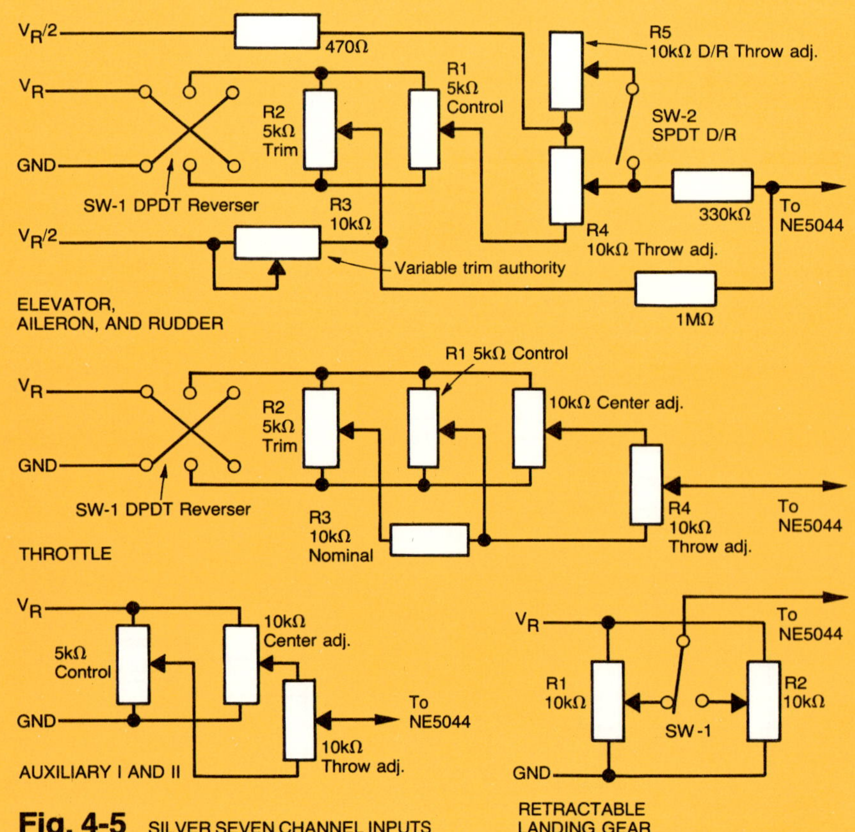

Fig. 4-5 SILVER SEVEN CHANNEL INPUTS

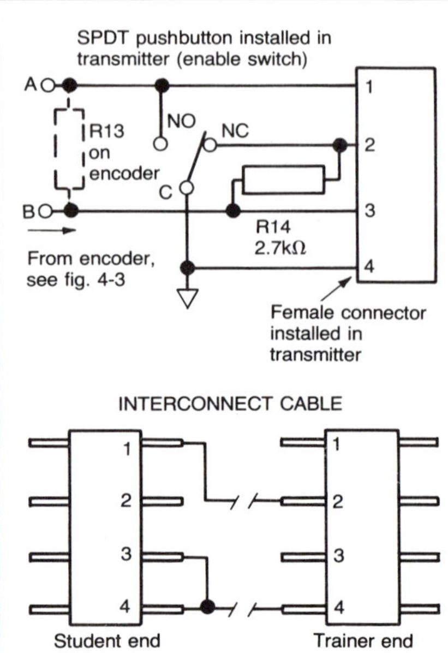

Fig. 4-6 SILVER SEVEN BUDDY BOX CONNECTION

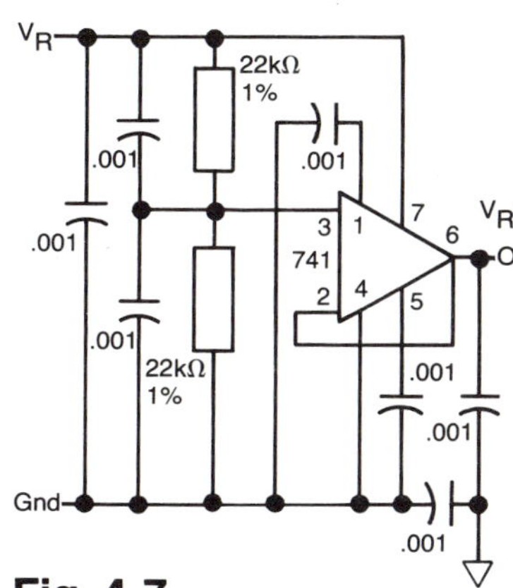

Fig. 4-7 VR/2 DIVIDER

SILVER SEVEN NONLINEAR AMPLIFIER MODULE
(one of three identical sections shown)

Fig. 4-8

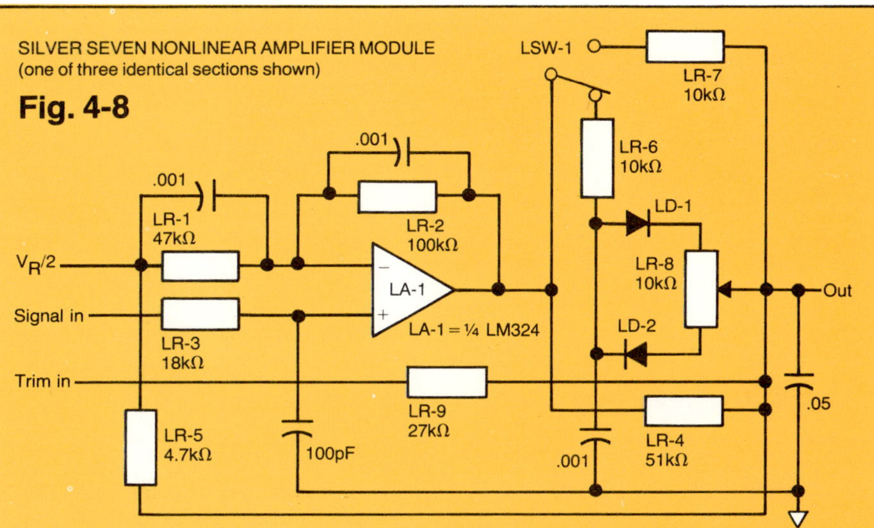

LR-4: Substitute 33kΩ-75kΩ to vary low slope.
LR-5: Substitute 3.9kΩ-10kΩ to vary both slopes simultaneously.
LR-6: Substitute 0-10kΩ to vary high slope.

Note: ¼ of LM324 not used.

changes minutely, $V_R/2$ will still hold true no matter what happens beyond the output. This provides a high-impedance output that prevents interaction between the channels. Further, any minute drift in V_R is automatically compensated.

The basic input circuitry for channels 1 through 4 is in Fig. 4-5. SW-1 is a DPDT switch set up to reverse the polarity of the supply voltage to the control and trim pots. R1 is the control pot on the stick and R2 is the trim pot. Both must be electronically centered. R1 has a broad adjustment tab that permits precise setting of the electronic center during system setup so that the servo doesn't shift position when reversed. Both R1 and R2 are free to travel over ± 30 degrees; R3 produces the desired ratio of authority for R2 such that trim will result in servo rotation range of +10 degrees to as much as 30 degrees while stick motion produces a nominal ± 45 degrees rotation. Setting R3 changes the ratio. In this way, the Silver Seven allows for variable trim authority. Variable trim is used as follows: A newly completed model may require considerable trim to correct its flight path, so trim might be set to yield, say, 25 degrees of travel. After the mechanical linkages are adjusted to trim the model, the 25-degree trim throw would be a nuisance, so trim would then be reduced to, say, 10 degrees for full trim tab movement. R5 and SW-2 permit the upper end of R4 to be paralleled, effectively moving the wiper closer to $V_R/2$ to reduce control throw to any desired level for dual-rate.

Remember that the control input voltages from R1 and R2 are at exactly $V_R/2$ when the stick is at neutral, and that as the stick is moved away from neutral, the voltage varies ± 0.5 volt. For channel 4, the center-adjust potentiometer is assumed to be at exactly $V_R/2$ also. (If deliberate offset is introduced, switch reversal results in a servo position switch.) Range-adjusting resistor R4 acts as a voltage divider; the closer the wiper is set toward $V_R/2$, the less the voltage swing to the NE5044 and vice versa. As noted earlier, if no deliberate offsets in center are planned, the rate-adjust may be connected to $V_R/2$ as for channels 1, 2, and 3. Since dual-rate isn't needed for throttle, the low-range pot SW-2 is deleted on that channel.

Retract gear signal-conditioning is simple. SW-1 selects between R1 or R2. One is set to give a control pulse width of 1.0 ms and the other is set to give 2.0 ms. Auxiliary functions are a simplification of the signal-conditioning for channels 1 through 3. There is no trim, no reversing, and no dual-rate. Otherwise the description above applies.

Channel 5 is routed directly to pin 5 of the NE5044, but the other six channels are routed via a "patch" section of the PC board that allows the kit-builder to route control functions for Mode I, Mode II, or whatever is desired. A 100-kilohm resistor in series with all seven inputs reduces potentiometer nonlinearity effects, and, with the 100-picofarad bypass capacitor, ensures that no RF enters the input side of the NE5044.

Nonlinear amplifier

Two optional plug-in modules, a nonlinear amplifier and a mixer, are available to the builder of the Silver Seven. Figure 4-8 is the schematic of the nonlinear amplifier module. It is based on an opamp and the circuit is duplicated for each nonlinear channel; the Silver Seven has three duplicate circuits for ailerons, elevator, and rudder. When the nonlinear module is used, the main PC board layout automatically routes the trim input to the module board.

The input from the desired channels enters the nonlinear amplifier via LR-3. LA-1, LR-1, LR-2, and LR-3 form an amplifier that increases the control input voltage to about 2.5 volts for full stick throw. That amplification is linear. With LSW-1 in the position shown, the amplified signal is applied to the parallel path formed by LR-4 and LD-1/LD-2. The forward breakdown voltage (the point at which the diode begins to conduct when plus voltage is applied to the anode) is about 0.5 volt. Thus, as long as the signal out of LA-1 is less than about 0.5 volt, 51-kilohm LR-4 forms the resistance to ground (in this case $V_R/2$) via LR-5. The resistance before breakdown of LD-1/LD-2 is in the megohm range and only about 10 percent of servo motion results from 20 percent of stick travel. LR-5 acts as a shunt to reduce the slope of the output from LA-1. When the input to LA-1 reaches about 0.1 volt, representing about 20 percent of stick travel, LA-1's output increases to about 0.5 volt, and LD-1 or LD-2 breaks down. The parallel resistance between LA-1 and LR-5 now becomes the inverse sum of LD-1 and LR-4 with the breakdown resistance of LD-1 equal to a few hundred ohms.

The output from LD-1 and LD-2 is impressed on LR-8, a pot that is used to equalize the broken-down resistance between LD-1 and LD-2 since such diodes are not noted for consistency. That path now contains a nominal 5-kilohm resistance (half of LR-8) in addition to the few hundred ohms for LD-1 or LD-2. Thus, if one assumes that LD-1, after breakdown, is 500 ohms and that the leg of the pot on that side is 4500 ohms for a total of 5000 ohms, the resistance between LA-1 and LR-5 becomes

$$\frac{1}{R} = \frac{1}{51{,}000} + \frac{1}{5000}$$

and the resistance is about 4500 ohms. The slope of the curve then becomes

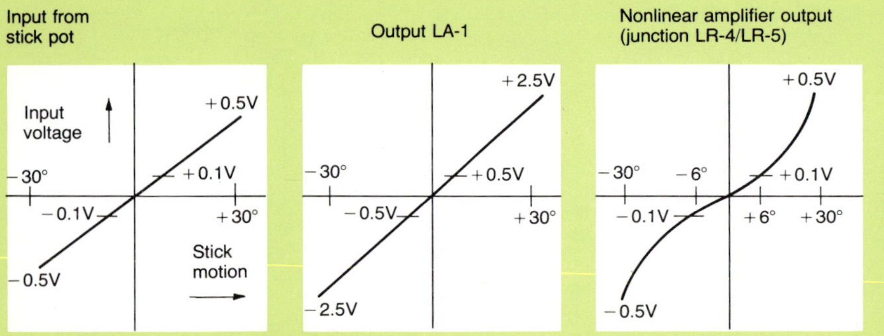

Fig. 4-9 INPUT AND OUTPUT SLOPES

much higher after input exceeds ± 0.1 volt and continues at that slope, for ideal diodes, to the end of stick travel. Again, remember that the function of LR-5 is to shunt the voltage so that the final output does not exceed ± 0.5 volt as for normal linear operation, Fig. 4-9. LR-4 may be varied from 15 kilohms to 51 kilohms via the plug-in sockets to change the slope of the lower curve. LR-6 is provided in series with LD-1/ LD-2 and LR-8 to permit the upper slope to be varied similarly to the lower slope. Thus you may tailor control sensitivity to suit your preference.

When the total resistance between LA-1 and LR-5 equals about 8 kilohms, the slope of the output is about 1.5 degrees of servo motion per degree of stick motion, the same as for normal linear operation. By including LSW-1 and LR-7 we can select, at will, either normal linear output or nonlinear (logarithmic) output since the diodes are removed from the parallel path and LR-7 is substituted. Because we're not changing the gain of LA-1, amplifier offset does not occur when the switch is made and there is no change in servo position. If trim were introduced along with Signal-In at LR-3, nonlinear operation would be adversely affected. This is avoided by adding trim after all nonlinear amplification has occurred. The .05-microfarad capacitor stabilizes the output and negates any spurious noise introduced by the amplifier. The remaining capacitors are RF bypass capacitors.

Mixer

Electronically, the plug-in mixer is more complex than the nonlinear amplifier. The schematic is in Fig. 4-10. The mixer will accept any two inputs and bidirectionally or unidirectionally mix one with the other. The ratio of mixing is independently variable. When MSW-3 is open, the mixer is unidirectional, only one channel mixes with the other. MSW-1/MSW-2, a DPDT switch, selects mix or no-mix. When MSW-1/MSW-2 is set for mix and MSW-3 is closed, the mixer is bidirectional, both channels intermix. At the Channel B input A1, MR-1, and MR-3 act as an inverting amplifier with relative gain set by MR-1. The output from A1 is coupled to a second inverting amplifier formed by A2, MR-5, and MR-9 set at unity gain. This second amplifier simply re-inverts the control voltage so that the servo moves in the same direction whether the mixer is switched in or out by MSW-2. MR-11 reduces the effect of control pot nonlinearity. You can visualize that if MR-1 were 51 kilohms, both amplifiers would produce unity gain and that the input and output level would be the same.

The output from A1 is also routed via MR-7 to the other half of the mixer formed by A3, A4, and related components. The Channel A input is impressed on A3. A3, MR-2, and MR-4 form another inverting amplifier with gain set by MR-2. The inverted output from A3 is routed via MR-8 to be summed with the output from A2 and the output from A2 is biased by the output from A3 to mix Channel A into Channel B. When MSW-3 is open, this mixing does not take place and the mixer is unidirectional. The output from A3 is also routed via MR-6 to A4. A4, MR-6, and MR-10 form another inverting amplifier whose gain can be varied to adjust for differences in servo travel.

Note that for bidirectional mixing any adjustment of MR-1 and/or MR-2 affects both servos. We want one output to be independently adjustable. How do we achieve this? The re-inverted output B→ A (the symbol means Channel A mixed into Channel B and vice versa) mixes outputs from A4 via MR-12. The gain for the Channel A side can be set to give less than unity or unity gain. You may ask why no-mix isn't obtained by opening the interconnect via MR-7 by another switch similar to MSW-3. This would avoid the complication of MSW-1/MSW-2. The answer lies in the constraints of mixing. With no-mix, each servo has a full range of ± 45 degrees. When bidirectional mixing is selected, each channel contributes ± 22.5 degrees, and can do that only when the output is half the input. Thus, full stick throw for Channel B produces ± 45 degrees on one or both servos, depending on setup. If we simply opened MR-7 and MR-8, the servos would have only ± 22.5 degrees

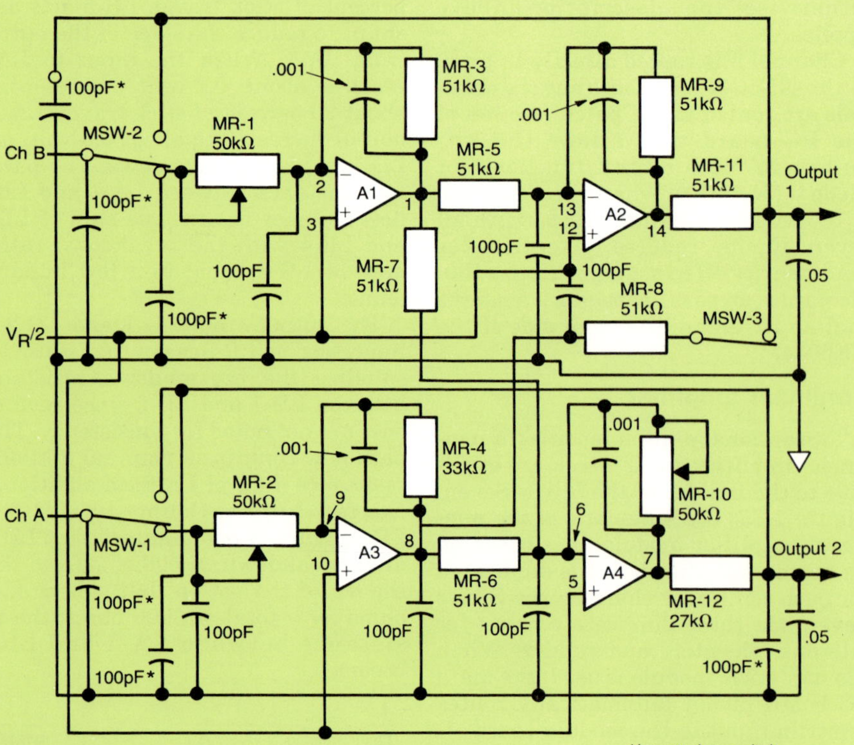

Fig. 4-10 SILVER SEVEN MIXER

*Located on switch, grounded to case

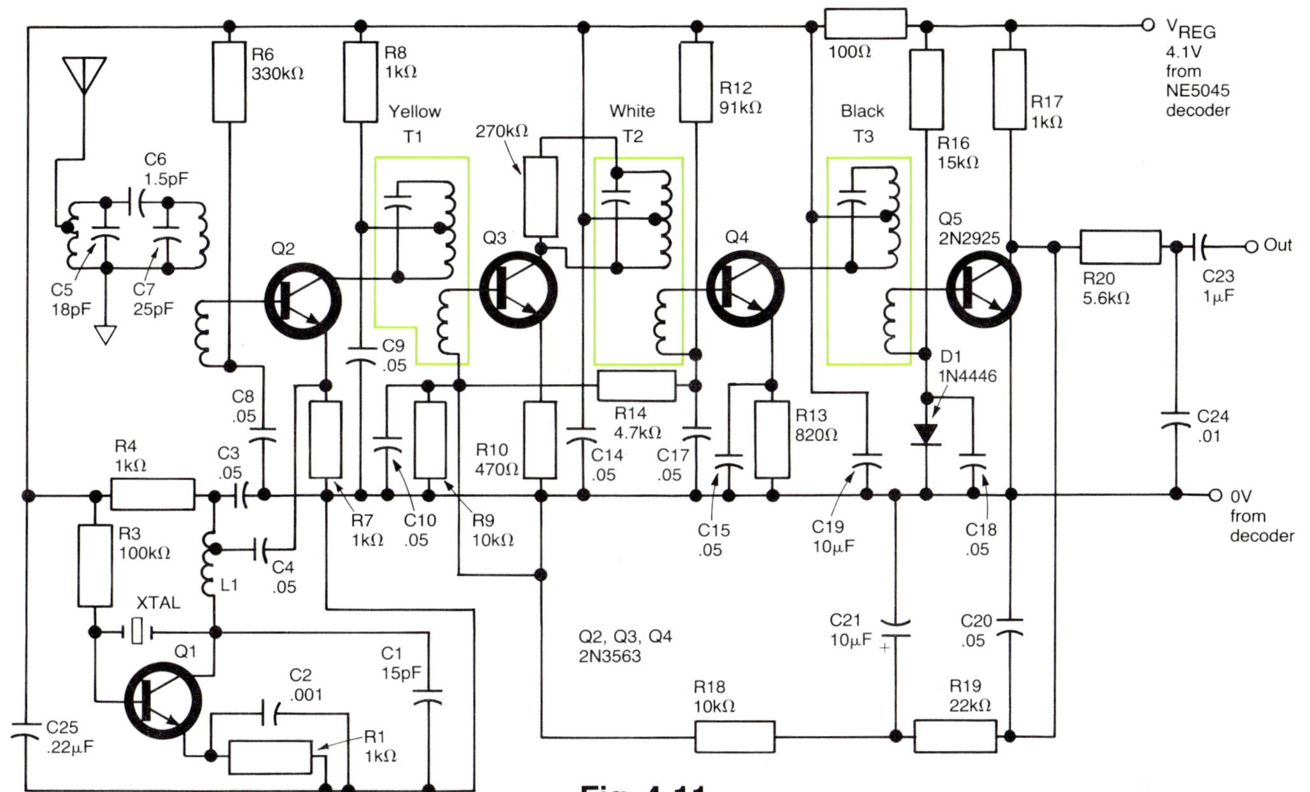

Fig. 4-11 SILVER SEVEN RECEIVER RF SECTION — DISCRETE COMPONENTS

throw from their respective sticks for no-mix. That would be unacceptable. If we move MSW-1 to no-mix, the servos have a full ± 45 degrees range.

When MSW-2 is closed for bidirectional mix, the inverted output from A1 is routed to A4 via MR-7. MR-7 and MR-6 have the same value, so the relative gain of A4 remains the same.

The capacitors in both the mixer and nonlinear amplifier modules have the sole function of bypassing to ground any RF picked up from the RF deck. This ensures that the circuits are not affected by RF.

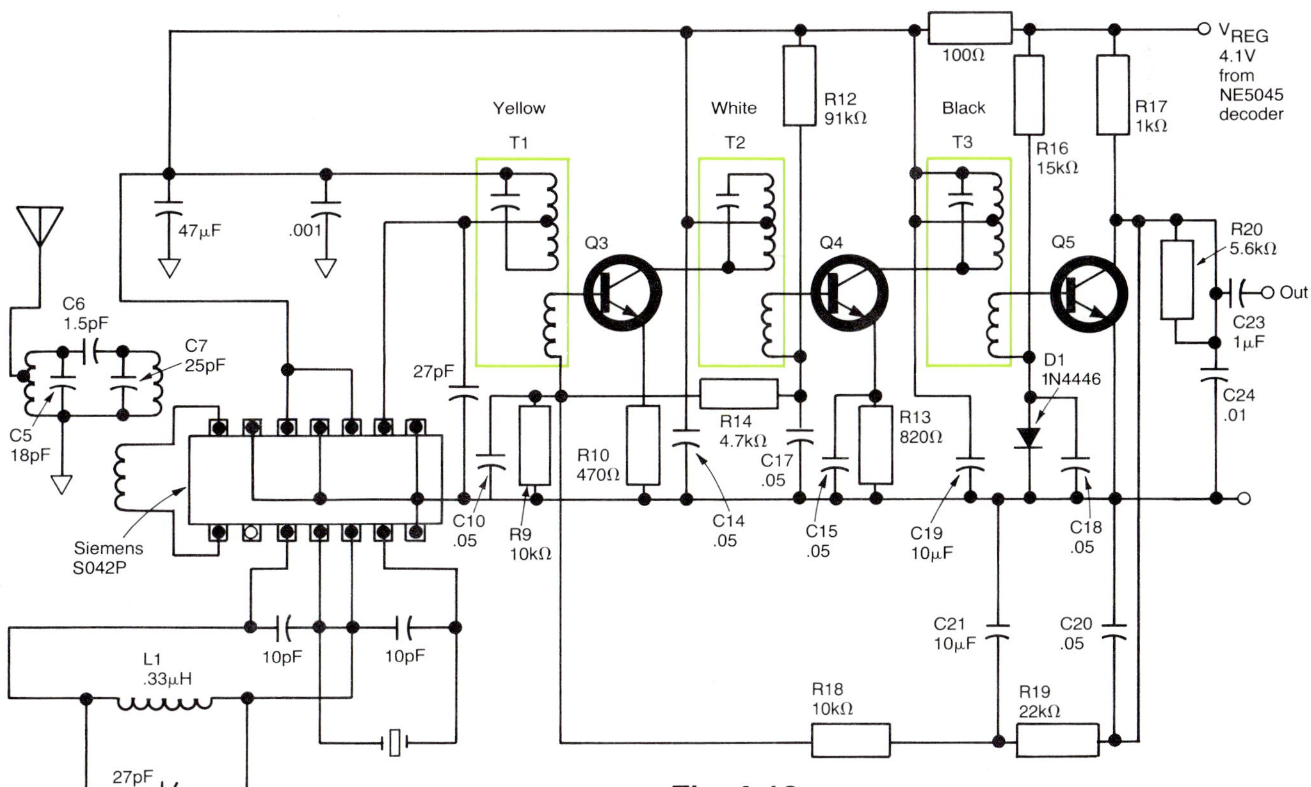

Fig. 4-12 SILVER SEVEN RECEIVER RF SECTION — IC

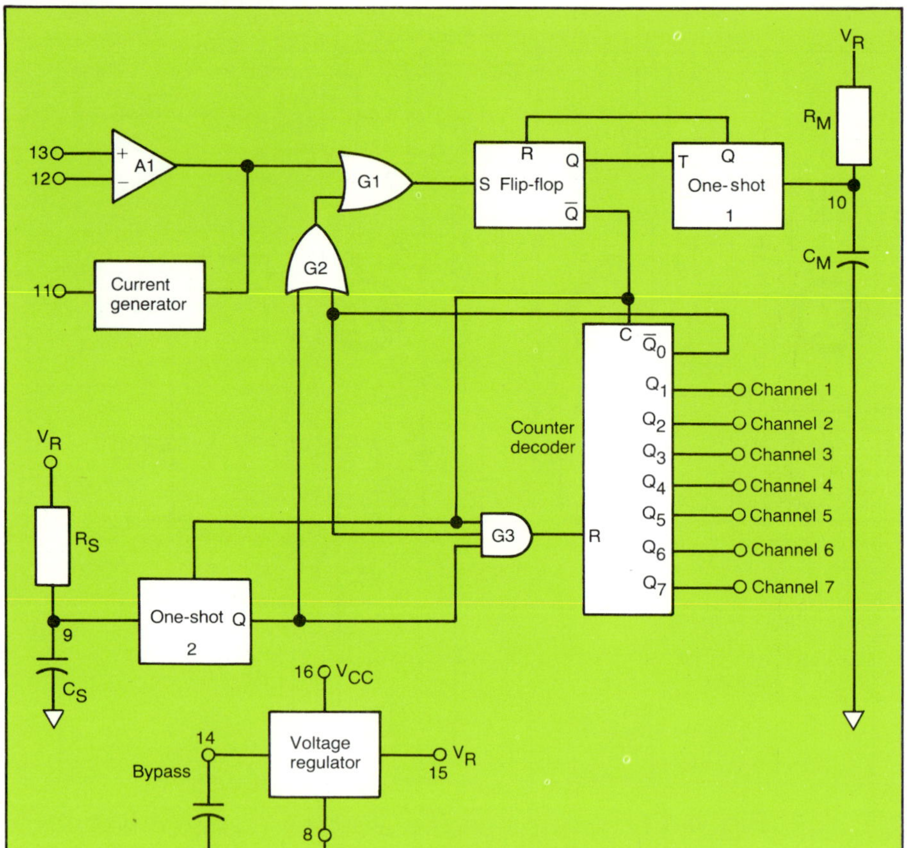

Fig. 4-13 NE5045 DECODER

Receiver

The Ace Silver Seven receiver consists of the RF section and the NE5045 decoder integrated circuit. The NE5045 provides 4.1 volts regulated to the RF section. Figure 4-11 is the schematic for the RF section built with discrete components; Fig. 4-12 shows the same RF section using the IC. The IC version is vastly simpler because many functions of voltage regulation and processing of the detected signal are performed by the NE5045.

The incoming signal is selectively passed by the double-tuned front end formed by L2, slug-tuned inductors L3/L4, and capacitors C5, C6, and C7. Q1, L1, R1, R3, R4, C1, C2, C3, C25, and the crystal form the local oscillator. Mixer transistor Q2 is biased into its linear operating region by R6. C4 couples the local oscillator signal to the emitter of Q2 where it is mixed with the incoming RF signal arriving at the base of Q2. The Q2 collector output is the complex signal described earlier, but has the desired 455-kHz IF. In the IC version, the mixer and oscillator functions are handled by the Siemens S042P IC double-balanced mixer. In both versions the 455-kHz IF envelope is the input to T1. Q3, Q4, T1, T2, T3, and related components form the IF strip. D1 is the detector diode. The output from Q5 is the detected positive-going pulse train that is a duplicate of the transmitted encoded signal. C10, C20, R12, and R19 are the automatic gain control circuit. The output from the RF section is an audio pulse train that varies in amplitude somewhat with differences in antenna orientation and range from the transmitter. The function of the AGC circuit is to sense this instantaneous change in output amplitude and adjust the gain of the IF stages to resist any such change in output pulse amplitude.

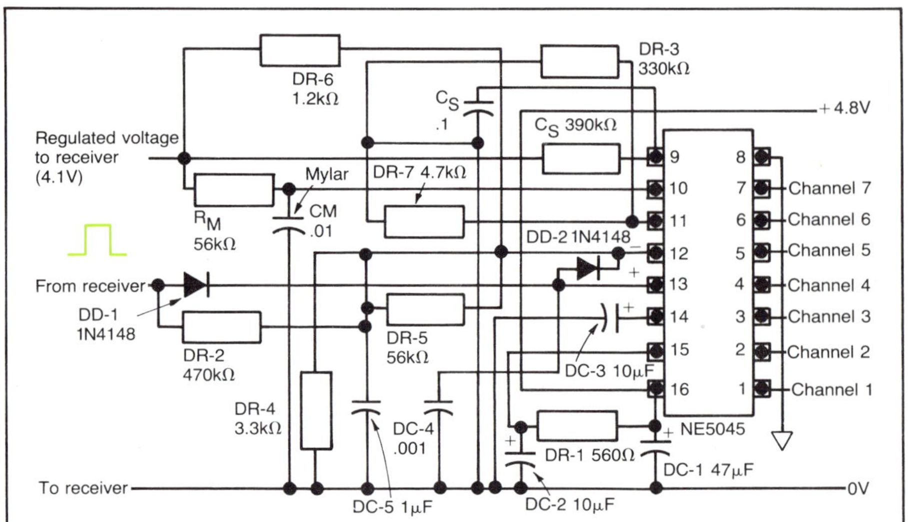

Fig. 4-14 SILVER SEVEN DECODER

Decoder

The NE5045 decoder is a serial-input, parallel-output integrated circuit containing all the necessary active circuitry to decode up to seven channels. It also provides regulated voltage for the RF section and for certain functions inside the chip. Figure 4-13 is the decoder's block diagram. Figure 4-14 shows the circuit used in the Silver Seven decoder. For convenience, I've cross-referenced the schematic values for passive timing components to the block diagram.

The feedback current generator provides positive feedback, creating hysteresis in the input switching levels. Hysteresis prevents false triggering due to noise or IF amplifier distortion. The receiver output is connected to the noninverting input, pin 13. In this case, the input threshold is set by the voltage difference between pin 12 and pin 13, established externally by the breakdown voltage of DD-2, DR-4, and DR-5.

The amplified signal from A1 is gated by G1 and in turn sets the flip-flop (FF). Assume that G2 is low. The combination of flip-flop and one-shot 1 produces a minimum pulse to clock the counter-decoder for each positive edge at pin 13 which exceeds the voltage on pin 12. The width of this pulse is $T_M = R_M C_M$. With this arrangement the system will not respond to any pulse after the first edge and before the end of T_M. In effect, the input is turned off for a period equal to T_M following the leading edge of each input pulse. The decoder's noise immunity is thus enhanced by the ratio of T_M to the period between input pulses. Obviously, T_M must be less than the shortest period between input pulses.

The counter is clocked each time the flip-flop is set and one-shot 2 is reset (capacitor CS is discharged). When the flip-flop is reset, C5 begins to charge through RS. The time constant $T_S = .85\ R_S C_S$ is normally much larger than the time between the input pulses, so

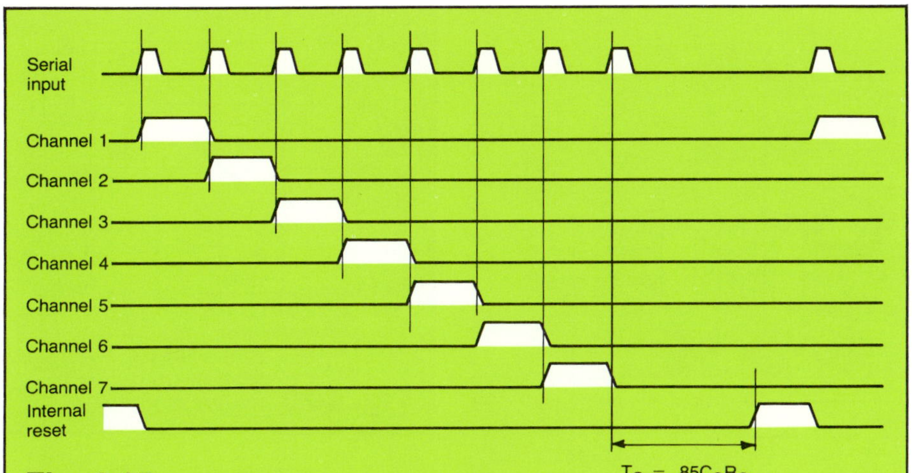

Fig. 4-15 SILVER SEVEN DECODER TIMING DIAGRAM

that the output of one-shot 2 remains low until the last pulse of a given frame is received. Figure 4-15 shows the decoder's timing diagram. After the last pulse in a frame (system synchronized), QO goes low and G2 goes high. The input is now disabled by G1 until one-shot 2 times out, at which time G2 goes low. This connection establishes synchronization in no more than one frame. It also prevents the counter-decoder from overflowing due to extra noise pulses in any given frame. Thus, any noise pulse in a frame will only affect those channels after the pulse and only in that frame.

If fewer than seven channels of input are used, then QO is high after the last pulse and the counter-decoder is reset when one-shot 2 goes high. Each channel has a totem pole output stage capable of sourcing 2 mA and sinking 1 mA, which is more than adequate for controlling contemporary servo amplifiers.

The voltage regulator operates in one of two modes depending on the power supply voltage. If V_{CC} is greater than 5 volts, the voltage regulator acts as a series-pass regulator with a nominal output of 4.1 V. When V_{CC} is less than 5 volts, the regulator acts as a dynamic decoupler where the bypass capacitor filters out line transients. The internal pass transistor acts like an emitter follower whose base is decoupled by the bypass capacitor. The value of capacitance depends on the degree of smoothing required and the amplitude of the line transients. The Silver Seven receiver-decoder will operate until V_R reaches 3.1 V.

Several components don't show in the block diagram, but are identified in the Silver Seven schematic. DR-1 is a pull-up resistor that ensures the regulator will start reliably. DD-1 and DR-2 form a clipping circuit that sets an input threshold of about 400 millivolts and clips all noise below that level. This, in combination with an electronically quiet receiver, extends useful range without the need to increase receiver sensitivity. DC-1, DC-2, and DC-3 are filter capacitors. DC-4 is an RF bypass. DC-5, DR-7, DR-6, and DR-3 help set the input threshold and provide further noise rejection and increased stability.

Servo amplifier

All contemporary RC systems use an IC servo amplifier, most often one made by Texas Instruments or Signetics. Hans Stellrecht and Robert Hamilton of Signetics, with considerable input from my own experience, designed the NE544 servo amplifier used in Ace R/C systems. Figure 4-16 presents the general block diagram for the NE544 in an operating servo. Figure 4-17 presents the detailed block diagram for the NE544 in the Silver Seven servos.

The input pulse, nominally 1.5 ms in duration, is first amplified, then fed to a pulse comparator. At the same time, the leading edge of the input pulse triggers the reference pulse one-shot. The lengths of the reference pulse and the incoming pulse are compared to establish the error pulse. Recall from the last chapter that the polarity of the error signal is plus or minus depending on whether the reference pulse is shorter or longer than the incoming control pulse. The error pulse is stored in a directional flip-flop. This polarity information is then gated by the Schmitt trigger output for a length of time proportional to the width of the

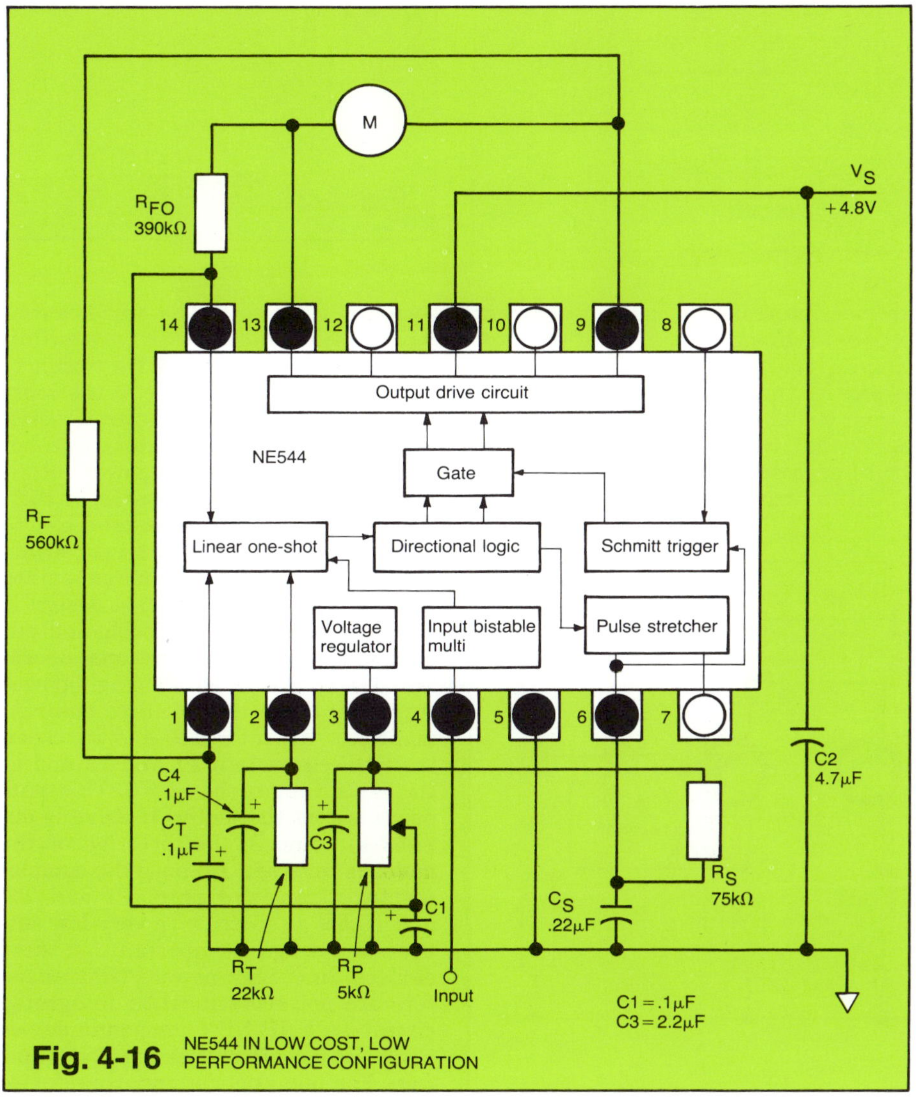

Fig. 4-16 NE544 IN LOW COST, LOW PERFORMANCE CONFIGURATION

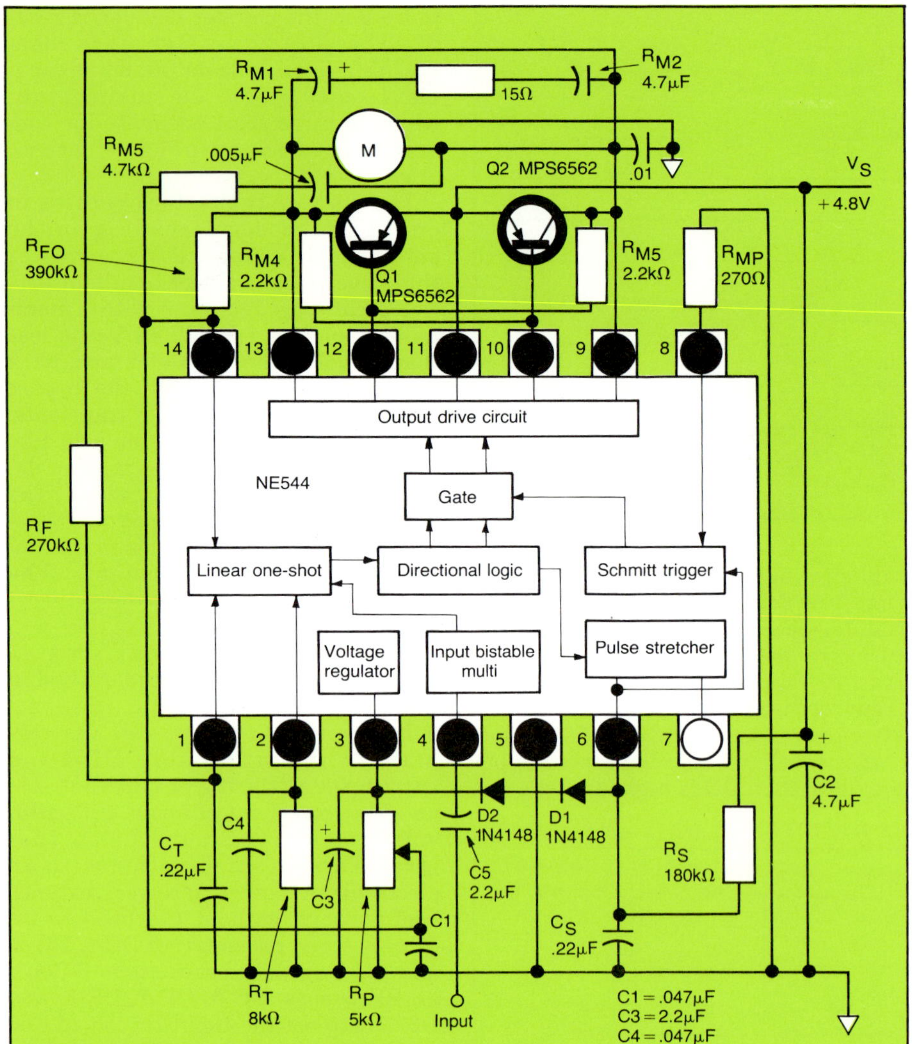

Fig. 4-17 SILVER SEVEN SERVO WITH NE544

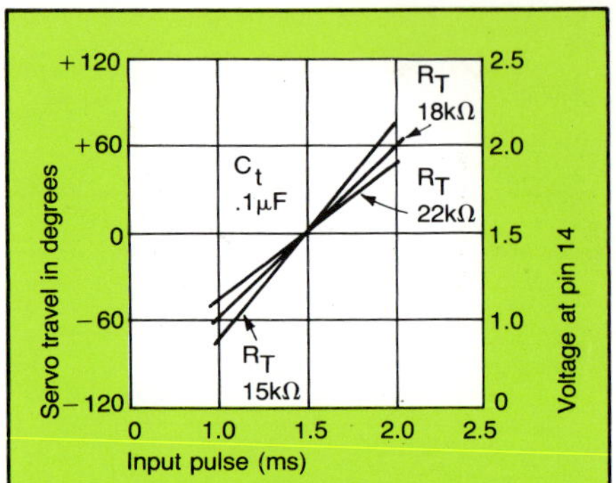

Fig. 4-20 SERVO OUTPUT TRAVEL ADJUSTMENT

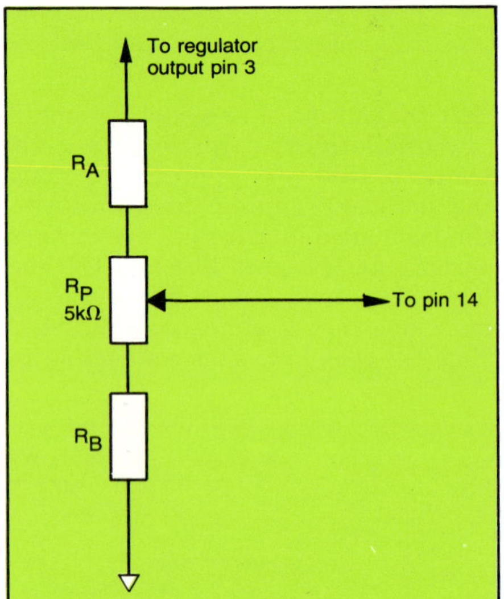

Fig. 4-21 INCREASED SERVO TRAVEL

error pulse. The output of the gate actuates a bidirectional motor drive circuit so that the servo motor can be driven in either direction, with the amount of drive proportional to the error in servo position. The feedback pot senses that error.

Two features of the NE544 are of special interest, the constant-current source for linear control and the motor drive circuit. The constant-current source for the NE544 is included in the linear one-shot and is in Fig. 4-18. Observe that the device is considerably more sophisticated than the original constant current arrangement discussed in Chapter 2. The NE544's configuration provides linearity to within 0.5 percent. The motor drive circuit is detailed in Fig. 4-19. One of the most difficult tasks in designing an IC servo amplifier is to prevent interaction between the relatively high power output stage and the low level control circuitry. This problem is compounded by the need to drive an inductive load which causes large voltage and current spikes while switching. The NE544 design is based on a conventional bridge output circuit that allows bidirectional operation of a DC motor with a single power supply. If drive is applied to point A, transistor T3 saturates and turns on the PNP equivalent T1 and T2 and at the same time turns on T5. This puts power across the motor. If drive is applied to point B, the direction of the motor is reversed. Because the complementary transistor pairs T1 and T2, and T5 act as switches, a very low saturation voltage is important for these devices. Since high-power PNP transistors are not economical in integrated circuits, a PNP-NPN compound device had to be used to realize the PNP function. For our applications, this device

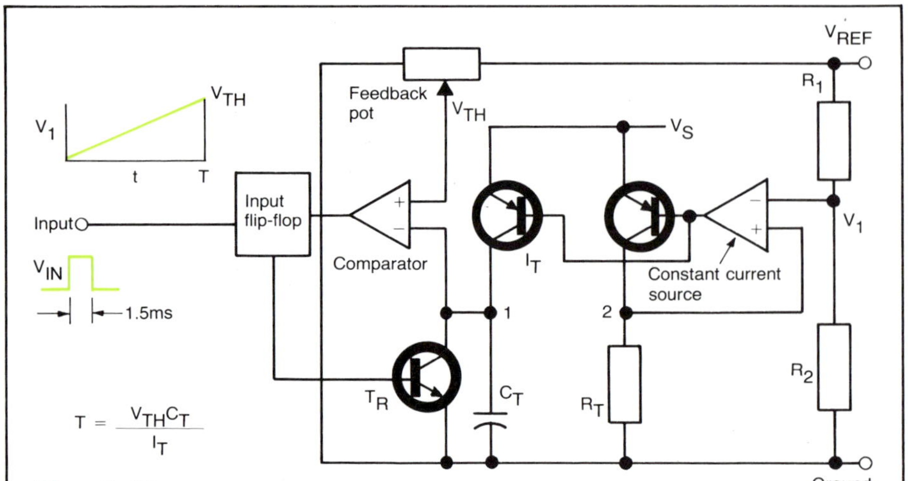

Fig. 4-18 CONSTANT CURRENT SOURCE

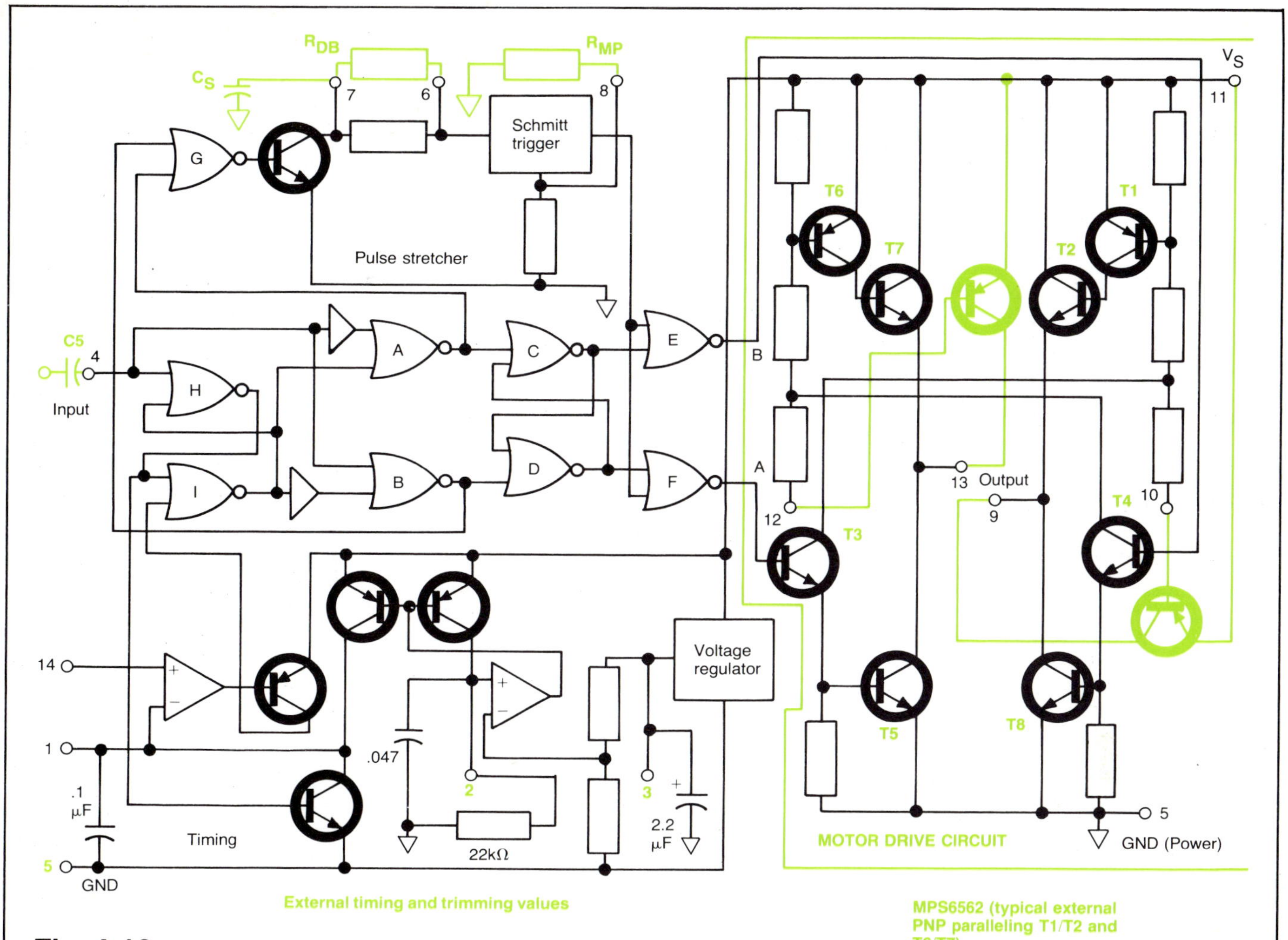

Fig. 4-19 SERVO MOTOR DRIVE CIRCUIT

has been bypassed with an external discrete power PNP, Fig. 4-17.

Figure 4-17 shows the full diagram for the Ace servo amplifier that uses linear feedback and external PNP transistors. For special applications, the kit-builder can change the rotational travel range of the servo output by selecting R_T in accordance with Fig. 4-20. If greatly increased travel is required, say 180 degrees for retractable landing gear, the circuit in Fig. 4-21 is appropriate. It can be accomplished by placing R_A and R_B in series with the end terminals of the feedback pot.

5 Kits

You may purchase kits for receivers, transmitters, servos, and accessories such as battery chargers, or you may purchase a kit for a complete system. In this chapter, I'll give general advice about component identification, soldering, and wiring. Then I'll describe how to build receivers, transmitters, servos, and battery packs.

Familiarization

Haste is the greatest enemy of the kit-builder. When you bring the kit home from the hobby shop or other retail outlet, or it arrives in the mail, carefully open the cardboard shipping carton and remove the assembly manual. Read the manual, several times if necessary, until you understand the assembly sequence from start to finish. Don't neglect the sections on theory of operation, because you'll need to know how the system works when the time comes for troubleshooting and repairs.

Parts for each assembly or subassembly are packed in separate paper or plastic bags. Unpack one bag at a time, keeping the contents of each bag together. Muffin tins and egg cartons make handy parts holders. Resistors, capacitors, and other components with long leads can be stuck in the open corrugations along the top edges of a small cardboard box. The components are reasonably sturdy but should not be dropped. If any parts are wrapped in foam plastic or metal foil, *do not unwrap them at this time.* The wrapping material protects these parts from damage by static electricity. Check the components against the parts list to make sure the kit is complete; if any parts are missing, write the manufacturer and he will supply them.

Component identification

If you are not familiar with electronic devices and coding methods, consult Fig. 5-1 and Table 5-1, and read the following descriptions.

The resistors used in RC systems are tubular and have leads protruding from both ends. The resistance is indicated by color bands around the body of the resistor. These are the first, second, third, and (sometimes) fourth colors. The fourth band (if present) is always silver or gold, while the first band is never silver or gold, so the first band is always easy to distinguish. If there is no fourth band, the resistor's tolerance is ± 20 percent of the nominal value; a silver band indicates 10 percent tolerance, while a gold band indicates 5 percent. The colors of the other three bands stand for the values in Table 5-1. The first band gives the first digit of the resistance value. The second band gives the second digit of the resistance value. The third band gives the multiplier for the value, that is, the number of zeros to be added. For example, the first band of a 270-ohm 10-percent resistor is red (2), the second is violet (7), and the third is brown (add 1 zero). The fourth band is silver for 10 percent tolerance. A 15,000-ohm resistor is brown (1), green (5), and orange (add 3 zeros). The following symbols are used on schematics and wiring diagrams: Ohm is the Greek letter omega, Ω; kilohm (1000 ohms) is kΩ or simply k; megohm (1,000,000 ohms) is MΩ or meg. Parts lists and wiring diagrams usually state the value of the resistor and follow this in parentheses with the color bands, as so: "270-Ω ¼-watt 10% resistor (red-violet-brown-silver)."

Disc capacitors have various markings. A disc capacitor will always have two, three, or four digits indicating the value of capacitance in microfarads (μF) or picofarads (pF). However, there is some inconsistency in the characters that follow the digits. A 10 pF, which is the same as 10 micromicro farads, may have only the number 10 printed on the body. A 250 pF may be marked "250Z," a 47 pF may be marked "47K," and a .05 microfarad may be marked only ".05." Your best bet is to consult the parts list and wiring diagrams. Note also that μF sometimes appears as "mf" and "mfd." Do be careful not to confuse the capacitor's voltage rating, which may be printed as "10V" or "10kV," with the capacitance rating.

The ratings of tubular tantalum capacitors are usually clearly marked although several codes are used. The decimal may be indicated by a letter such as "R" or the decimal may be omitted. The first two digits may be followed by a third number indicating the number of zeros to be added. Usually you need only read the first two digits: Thus, a 4.7-μF tubular tan-

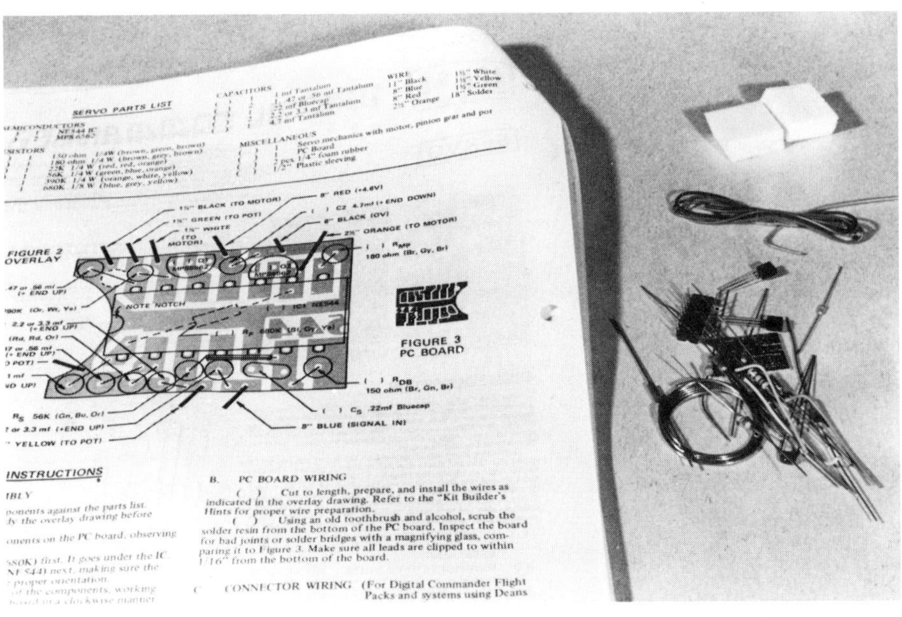

Color	1st Digit	2nd Digit	Multiplier
Black	0	0	1
Brown	1	1	10
Red	2	2	100
Orange	3	3	1000
Yellow	4	4	10,000
Green	5	5	100,000
Blue	6	6	1,000,000
Violet	7	7	10,000,000
Gray	8	8	100,000,000
White	9	9	1,000,000,000
Gold	—	—	5% tolerance
Silver	—	—	10% tolerance
None	—	—	20% tolerance

Table 5-1 RESISTOR COLOR CODE

Study the assembly manual before beginning to assemble the kit. Become familiar with each step-by-step sequence.

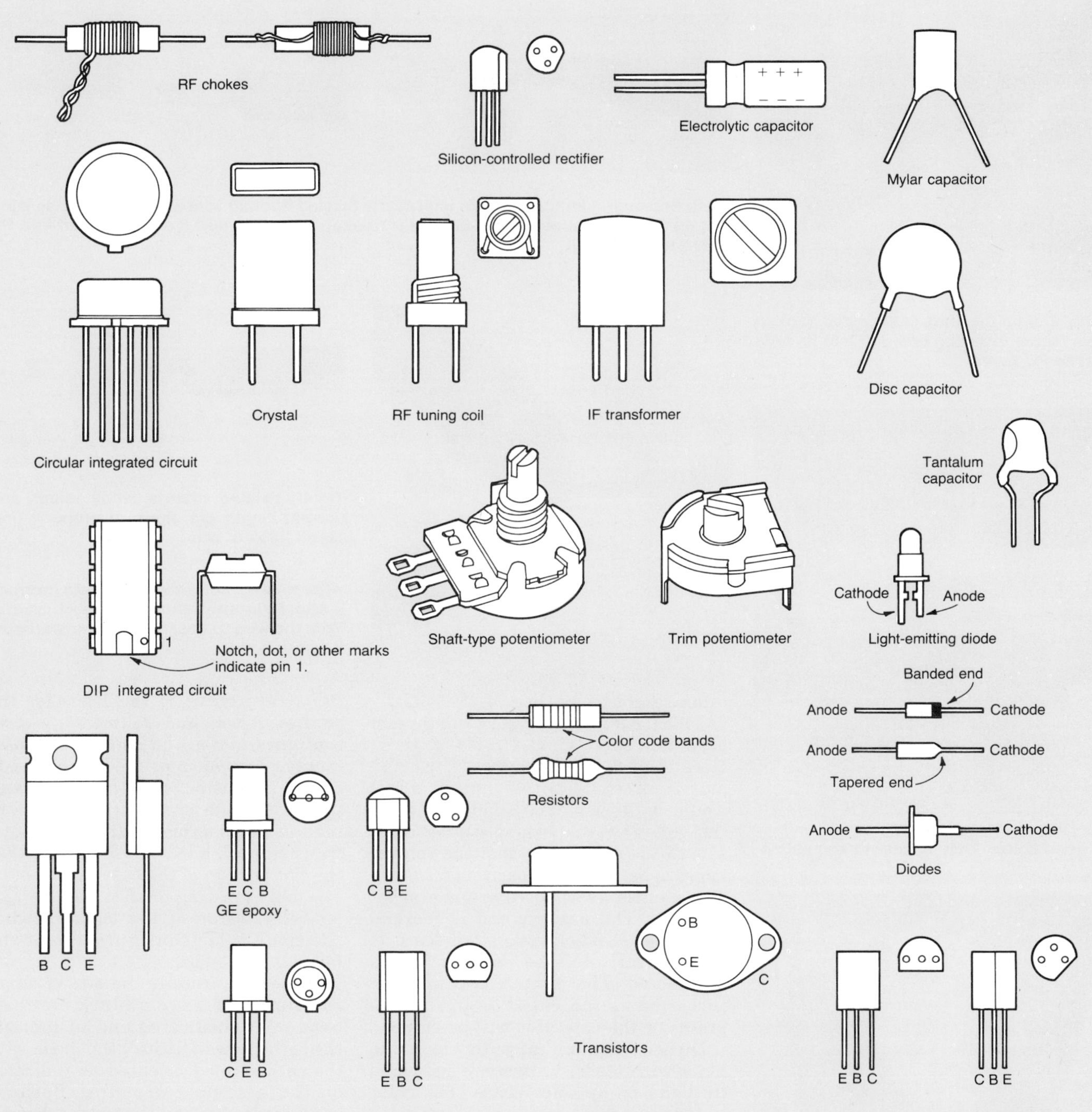

Fig. 5-1 ELECTRONIC COMPONENTS

Unpack the parts for each subassembly and arrange them neatly in muffin tins or egg cartons.

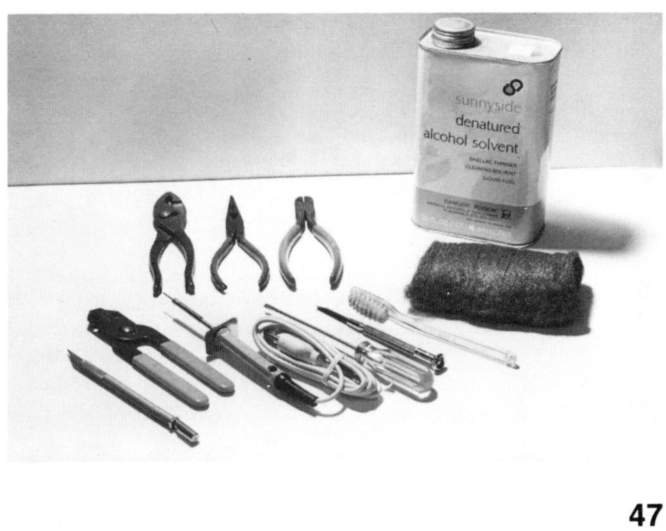

Kit-building requires only a few tools and cleaning supplies. Buy the best tools you can afford.

Twist and tin the ends of stranded hookup wire before inserting one end into its hole on the PC board.

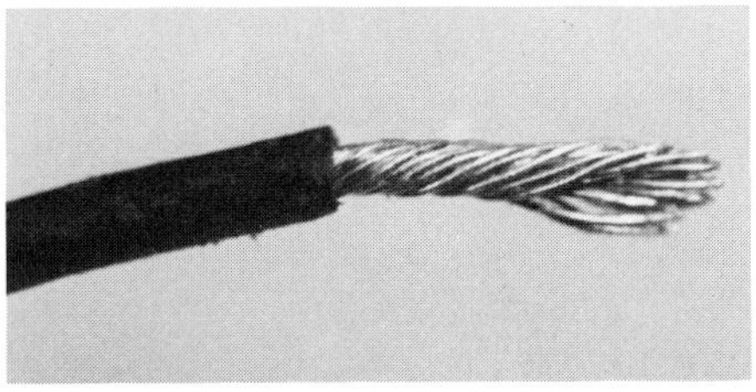

(Left) Protruding strands on this improperly twisted hookup wire will cause trouble later. (Right) Correctly twisted strands on this hookup wire will make it easy to insert into PC board after tinning.

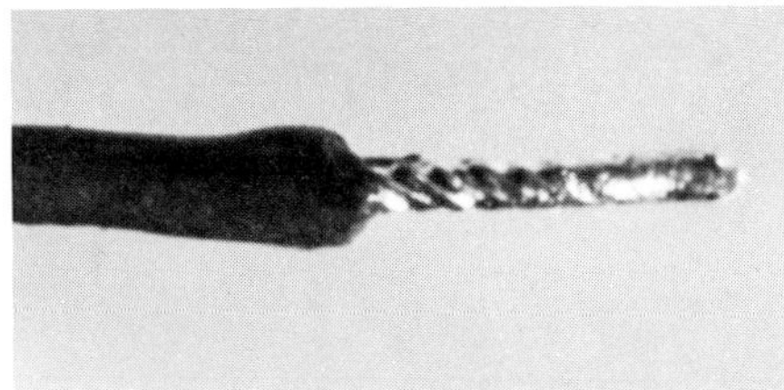

Tightly twisted strands, shiny solder, and correct length are signs of properly prepared hookup wire.

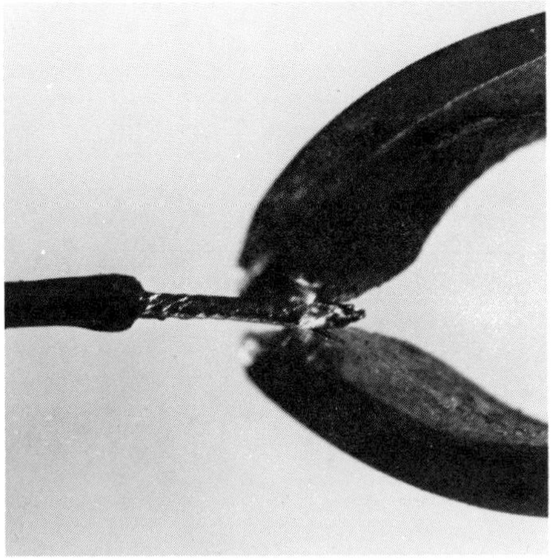

After tinning, cut the wire so it will protrude a short distance through the PC board. Trim the wire to final length after soldering it to the board.

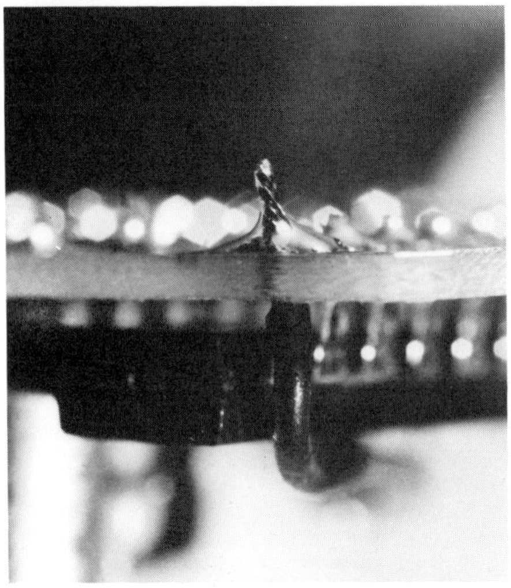

A good solder joint is smooth and shiny.

After solder has cooled, clip each lead to within 1/16″ or 3/64″ of the PC board.

talum might be marked "4.7M," "4R7," or "475." An alternative coding system looks like this: "CNS 475 B 050 K 1." Here the "475" stands for 4.7 μF (47 plus 5 zeros). "476" (47 plus 6 zeros) would be 47 μF. The "050" indicates the voltage. This sequence may occasionally be reversed so that the voltage appears before the capacitance rating. Tantalums have a positive and a negative end. The positive end is marked with a red band or some other feature, or, alternatively, the positive end is shouldered. The positive end must be installed as indicated on parts diagrams or the capacitor will be ruined.

Dipped tantalum capacitors look like blobs with leads. Values are indicated in digits or by color codes. The color code is the same as for resistors, with the top color the first significant digit, the middle the second digit, and the third the number of zeros to be added. The unit is always picofarads. Thus, the colors yellow-violet-green represent 47 plus 5 zeros or 4.7 μF. Yellow-violet-blue stands for 47 plus 6 zeros or 47 μF. The dot or "+" indicates the positive lead.

Chokes and other coils are wire-wound and may have exposed windings or dipped windings. If they are coded, the code is the same as for resistors except that the unit of measurement is microhenrys (μH) instead of ohms. If they are not coded, the parts list and assembly diagrams will clearly distinguish one from another, perhaps by pointing out a dot of color or a colored lead.

Transistors vary in size, shape, and the arrangement of the leads for the emitter, base, and collector. Several configurations are in Fig. 5-1. The part number is always printed on the body of the transistor, sometimes so small that you need a magnifying glass to read it, but it's there. When installing transistors on a PC board, be sure that the flat portion of the case matches the assembly drawing or that the triangular relationship of the leads matches the triangular configuration of the holes in the board.

Diodes are usually the size of small resistors and have a black or white band or a shouldered end to indicate the cathode lead. Install the diode with the cathode end oriented as indicated on the assembly drawing. Because most diodes are made of glass, handle them carefully. Light-emitting diodes come in many sizes and colors. The cathode is the longer lead, the colored lead, or the lead marked with a spot or notch in the case. If in doubt, insert the LED into the circuit: If it doesn't light, reverse the leads and try again. LEDs have plastic cases and lenses, so do not apply much heat when soldering.

Integrated circuits are mounted on circular, or more often rectangular, cases with a number of pins protruding. Pin 1 is marked by a notch or indentation in the case, or, for circular ICs, by a tab. Position these as shown in the assembly drawing. Some ICs, particularly CMOS devices, are ruined if exposed to static electricity before they are installed in the circuit. Their leads will be protected by a piece of grounding foam plastic or metal foil —

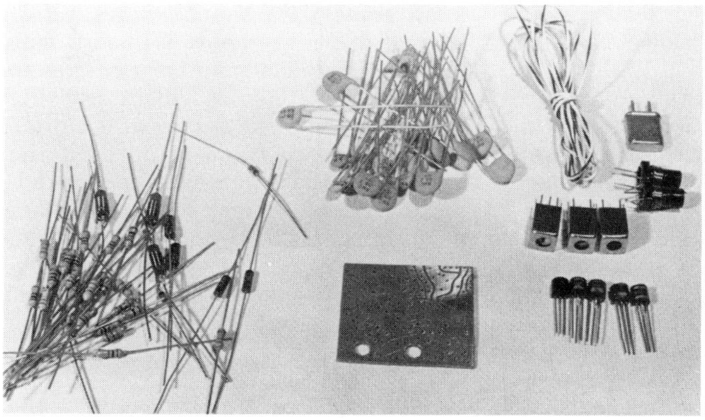

The receiver RF deck consists of a PC board, five transistors, two RF coils, three IF transformers, a crystal, several chokes and diodes, and many resistors and capacitors. Mount each component as called for in the assembly manual.

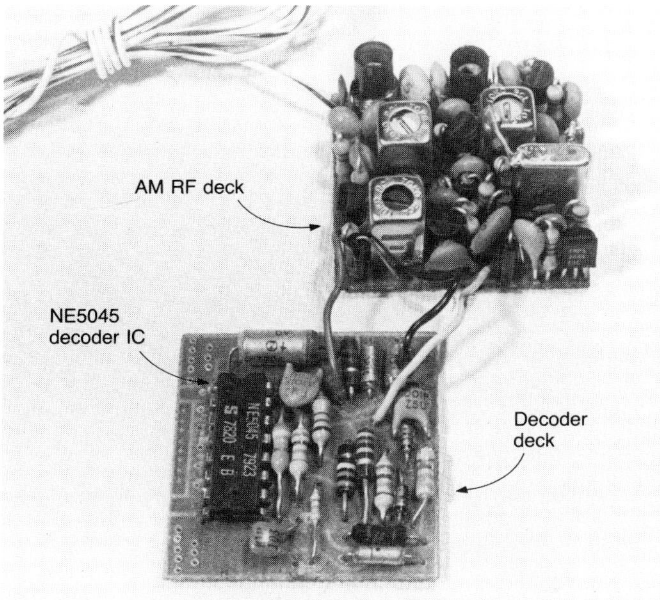

Signetics NE5045 IC in the decoder reduces the number of discrete components required. Two-deck design allows for eventual replacement of the AM RF deck with an FM RF deck.

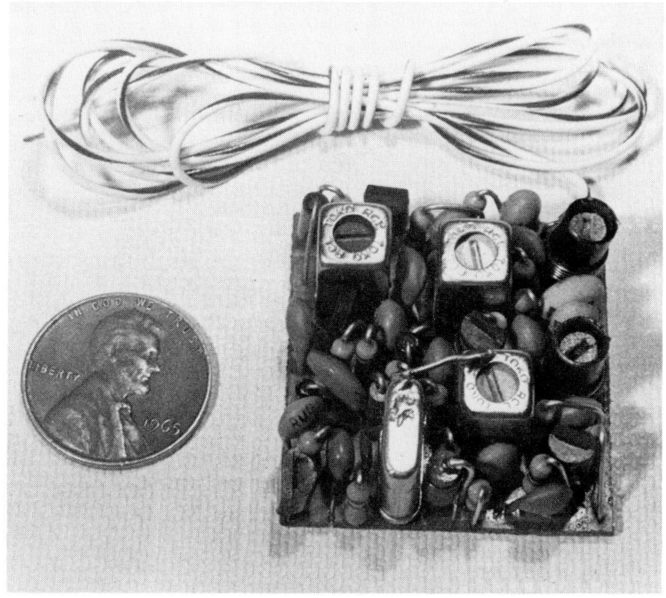

Completed receiver RF deck reveals that small size is made possible partly because many components are mounted vertically. Note how the leads of vertically mounted components are bent to a gentle radius.

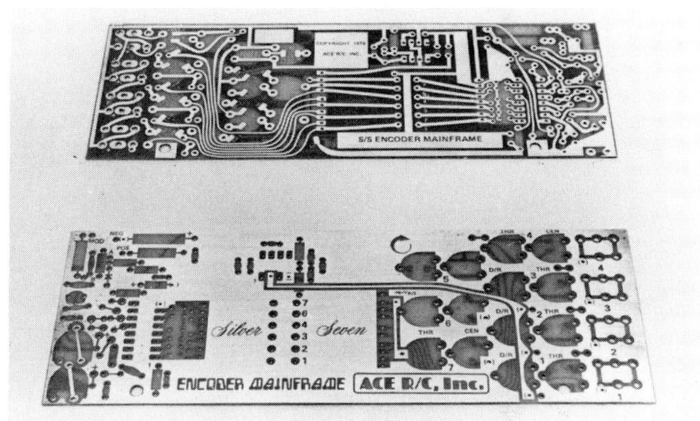

Silver Seven encoder uses a double-sided PC board. Scrub both sides with steel wool wetted with denatured alcohol before beginning to mount components.

do not remove this material until just before installing the IC on the PC board. Wear cotton clothes and be careful not to shuffle across a rug or otherwise generate static electricity when handling CMOS devices.

Component tests

Before beginning to assemble the kit, test as many components as possible. Most components (except ICs, transistors, crystals, and some capacitors) can be checked with an ohmmeter. Take the time to check components now and you'll be less likely to have to troubleshoot a nonfunctioning system later.

Check resistors with an ohmmeter and replace any that exceed their tolerance. Note that a 10-kΩ resistor with a silver band (10 percent tolerance) is good if the resistance ranges anywhere from 9 kΩ to 11 kΩ. Check diodes for shorts or other internal defects by touching the cathode end lead to the positive lead of the ohmmeter. Touch the negative ohmmeter lead to the other end, the anode, of the diode. Resistance should be very high — almost infinite. Switch the positive and negative leads and resistance should be no more than a few hundred ohms. LEDs are tested in the same way as other diodes, or you can use a battery in series with a potentiometer, Fig. 5-2. Chokes and coils can be tested for opens with an ohmmeter: A reading of infinity indicates a defective coil. Perform the same test on all windings of the IF transformers. Test potentiometers with the ohmmeter using the sequence in Fig. 5-3. Then check the open elements in the pots used in the servos or servo assemblies for roughness and irregular movement. Switches should make and break contact smoothly and firmly. Use the ohmmeter to be sure contacts open and close properly.

Accurate checks of capacitance require a capacitance meter, which most hobbyists do not own, so you'll have to trust the ratings of smaller capacitors. However, electrolytic capacitors, including tantalums, can be rough-checked with an ohmmeter. An electrolytic capacitor is good if the ohmmeter's needle flicks momentarily when the ohmmeter leads touch the capacitor leads. This gives no indication of value, but it's a good go/no-go test.

Check servo motors by placing 4.8 volts across the motor each way. Speed should be equal in both directions. Nickel-cadmium batteries should show 0.8 to 1.2 volts under no load even if they have been stored for a long time. Transistors are best checked for opens, shorts, and leakages with an inexpensive transistor checker such as the Heathkit IT-3127.

Crystals and ICs can't be checked with the equipment available to most

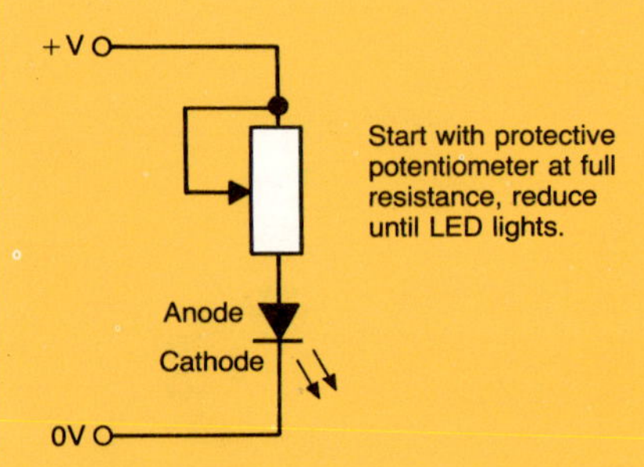

Fig. 5-2 TESTING LEDs

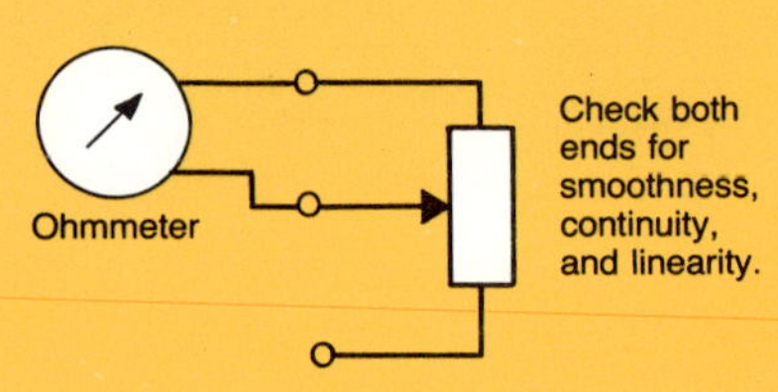

Fig. 5-3 TESTING POTENTIOMETERS

hobbyists, but have been tested by the manufacturer, so will almost certainly be good.

If your tests show that any component is defective, ask the manufacturer to send a replacement. Do not substitute components from surplus stores or grab bag selections unless you know they are not manufacturer's rejects.

Tools and cleaning supplies

For kit assembly, you'll need small diagonal-cutting and needle-nose pliers, a simple wire stripper, pocketknife or hobby knife, small screwdrivers, adjustable pliers, and a 25- to 37½-watt soldering iron with a tip not larger than 1/16″ in diameter. Ungar handle No. 777, with a No. 1235 37½-watt element, and a No. PL340 3/64″ spade tip is a good combination. Any similar iron will work well. Do not use a pistol type soldering gun: its tip is too large and the energy radiated by its transformer can damage delicate components. Also gather up a few pads of 000 or 0000 steel wool, denatured alcohol, and an old toothbrush.

You're now ready to begin assembling the kit. Assembly falls into three main stages: PC board wiring, interwiring, and alignment. PC board wiring involves soldering components to the printed circuit board, interwiring means connecting wires and cables from the board to other parts of the assembly, and alignment means adjusting transformers, coils, and other parts to tune the system. In addition, the parts of the system must be mechanically assembled with nuts, sheet metal screws, and other fasteners.

Receiver and decoder assembly

Begin construction by cleaning the receiver and decoder PC boards to remove any dirt and oxidation that would interfere with soldering. Scrub the boards lightly with steel wool wetted with denatured alcohol. Follow the assembly manual's step-by-step sequence exactly and insert the leads of the components in the appropriate holes. Mount all components except transistors tightly against the board, bending the leads of all components slightly outward to hold them in place when the board is turned over for soldering. Leave ⅛″ between transistors and the board to avoid overheating. As each

Top side of the encoder PC board identifies all major components. Mount switches and potentiometers first, progressing from the right end of the board to the left.

Two brackets (circled) on the lower edge of the PC board mount the completed encoder to the transmitter case.

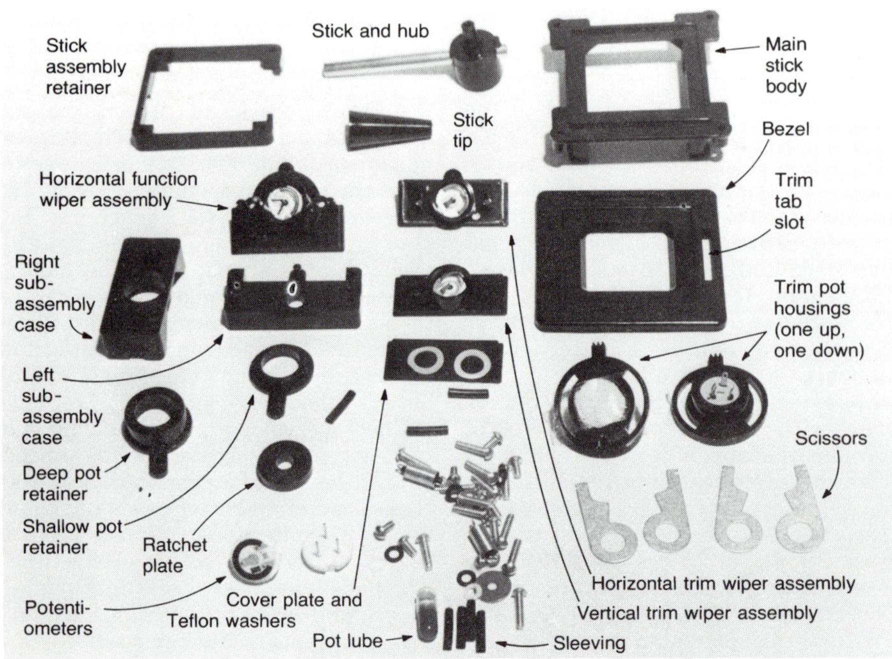

These are the parts for a D&R two-axis stick. Sticks are also available completely assembled and can save several hours' work.

Springs ensure that the control stick automatically returns to center.

part is mounted, check off that step with a pencil. Do not deviate from the step-by-step process; it has been carefully thought out to prevent parts from becoming inaccessible in subsequent steps. Even I get into trouble when I depart from the assembly sequence for the Ace Silver Seven, and I designed the kit!

After several components have been mounted, and when called for in the assembly manual, solder the leads to the board. Here's how:

Step 1. Wipe the tip of the hot iron with a damp sponge until it is shiny.

Step 2. Hold the tip of the soldering iron so that it touches both the lead and the circuit board foil.

Step 3. Place the solder between the iron and the joint being soldered so that it touches both.

Step 4. Remove the solder as soon as it melts and flows onto the lead and foil. Remove the iron.

Step 5. Allow the solder to cool. Inspect the joint. A good solder joint is smooth and shiny. A bad joint is dull, cracked, and lumpy. Bad joints most often result from not heating the joint long enough before applying solder, removing the solder too rapidly, or moving the joint before the solder cools. Reheat a bad joint and it will almost certainly turn into a good joint.

Step 6. Clip the leads to within 3/64″ of the board using small diagonal-cutting pliers or a toenail clipper. Hold onto the piece being cut off to prevent the cut-off leads from flying into your face.

Use only enough solder to form a strong joint — too much increases the possibility of solder bridges between the PC lands. Excess solder will also fill in the holes for wires that will be added later. You can clear a solder bridge by heating the connection and sucking up the solder with a braided copper wire called a solder wick (one brand is "Wik-It") or by using a Teflon-tipped syringe. Both of these types of solder removers are sold in all electronics stores. Clear holes by heating the solder, then push a scrap resistor lead into the hole. Withdraw the scrap lead before the solder hardens.

Use the solder provided by the manufacturer. Should you need more (though the quantity provided is usually more than enough), buy 18- or 24-gauge resin-core solder that is 60 percent tin and 40 percent lead. Ersin Multicore is excellent, as is Kester. Do not use acid-core solder. Acid-core solder leaves a flux residue that corrodes parts and voids all warranties.

Mount components on the PC board until all are in place. Be especially careful when mounting the crystal. As you might suspect, the crystal is fragile. Don't drop it. Some receivers require that a ground wire or brace be soldered to the crystal case. If so, be careful not to melt the blob of solder near the top of the case of the crystal — it seals a hole where the case was evacuated.

Secure the receiver front end tuning coils to the PC board with 5-minute epoxy or an alpha cyanoacrylate cement such as Hot Stuff, Jet, or Zap. When all components have been mounted, clean off any flux residue with a toothbrush wetted with denatured alcohol. Do not let the alcohol run into the IF transformers or any other components because the alcohol might dissolve glue joints or insulation in these parts. Now inspect the board with a magnifying glass or jeweler's loupe. Check for bad solder joints and solder bridges between lands. Fix any that you find. It's a good idea to have a friend inspect your work at this time — he'll often find mistakes that you've overlooked.

Interwiring

Wiring between the PC board and other assemblies should be neat and orderly. Route the wires or groups of wires as shown on the assembly drawings. Power supply wires are red (positive) and black (negative). Pairs or triplets of wires are usually different colors to aid in circuit tracing. Always use the color of wire specified in the directions. Twisted pairs or triplets of wire must be twisted before cutting to final length, because the twisting process shortens the length by up to 25 percent. Pairs are best made by gripping one end of two equal lengths of wire in a vise, chucking the other end in a hand drill, and using the drill to twist the wires. Be careful not to kink or break the wires. Triplets are best twisted by hand.

The leads of solid wire components are tinned (coated with a small amount of solder) during manufacture, but you must tin the ends of all stranded wires yourself. Strip off 1/8″ of insulation, twist the stranded wires together, heat the twisted strands with the soldering iron, apply a small amount of solder, and flip off any excess solder before it solidifies.

The Silver Seven and most other kit systems use Deans-brand connectors. Figure 5-4 shows how to wire these. Figure 5-5 illustrates typical wiring harness assembly techniques. Figure 5-6 shows how switches are wired. When preparing harnesses and attaching wires to connectors, be certain that all wires from battery packs have female connections. If you do otherwise, the pins on the male plugs will inevitably short out on some metal around the workbench and burn up the wiring harness. Use the polyvinyl chloride (PVC) sleeving supplied with the Deans connectors on all plug connections. Use rubber grommets to protect the servo and power leads where they

Trim potentiometers attach to the sides of the stick assembly. Use Teflon washers and tighten the bolts just enough that the trims move smoothly.

Wiring harnesses must be neatly twisted, carefully positioned, and firmly secured with cable ties.

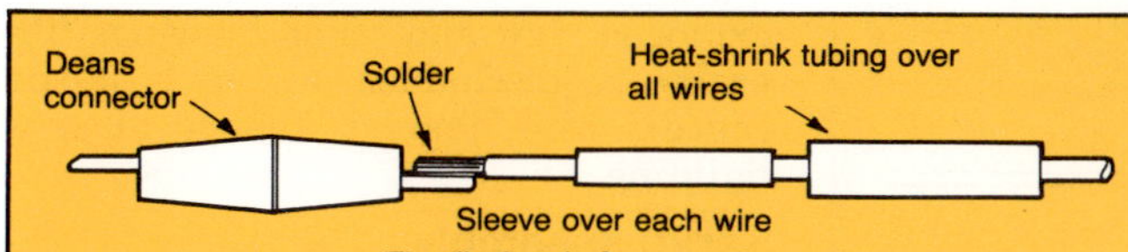

Fig. 5-4 DEANS CONNECTOR WIRING

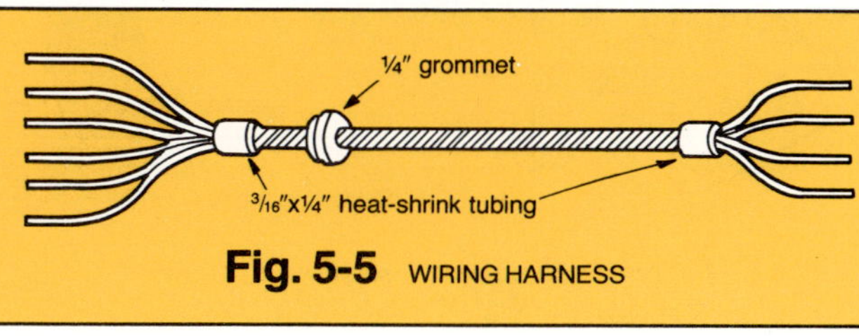

Fig. 5-5 WIRING HARNESS

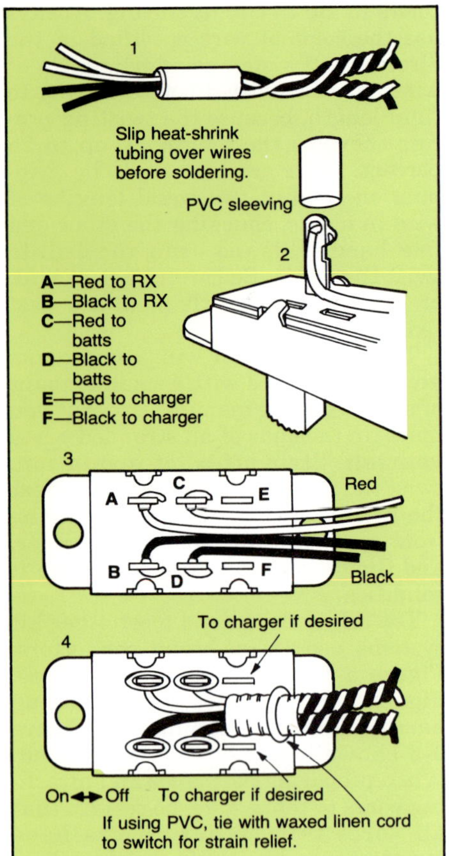

Fig. 5-6 TYPICAL SWITCH WIRING

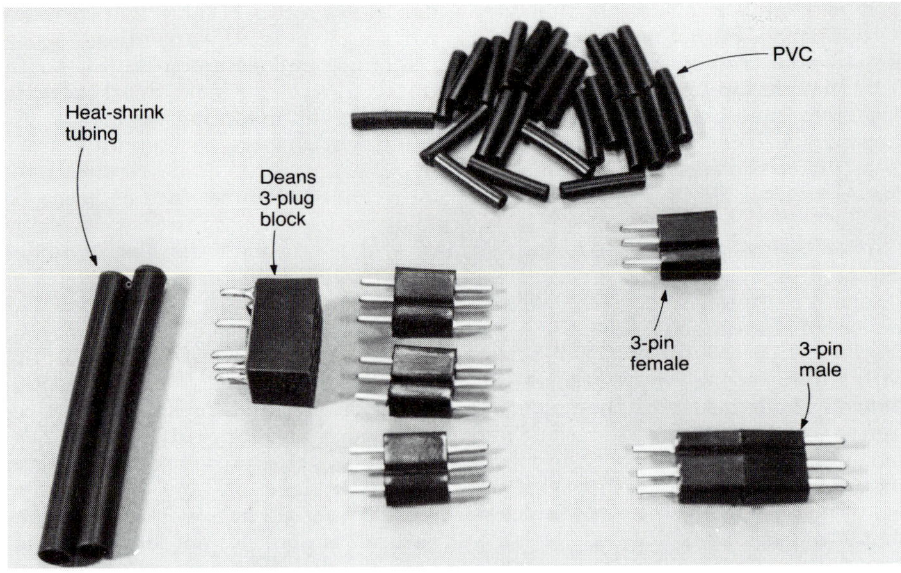

Deans-brand connectors and plug blocks are popular with RC modelers because they are sturdy and reliable. All connections must be insulated with PVC sleeving or heat-shrink tubing included with the kit.

exit the receiver case. I usually solder a cable to the receiver PC board first, slip it through the grommet, place sleeving over each wire, and then solder each wire to the appropriate connector pin.

When soldering stranded wire to connectors, first fasten the connectors together to minimize the possibility of wiring errors. Slip the tinned wire through the appropriate sleeving, heat-shrink tubing, and grommet; double-check to see that the correct wire goes to the correct terminal; and solder. Work quickly to avoid melting the plastic body of the connector.

When interwiring to moving parts, leave sufficient slack for the wires to flex. Attach all such wires to the movable part with nylon straps or waxed linen cord — do not permit the ends of the wire to take the strain of the motion. In addition, heat-shrink plastic tubing or PVC sleeving, or both, must be used where wires attach to switches and control pots. The only exceptions to these practices anywhere in kit construction are wires attached to D&R and similar transmitter control trim assemblies. These have no tie points for straps or cord, but because motion is limited little strain is involved. Sleeving and careful routing of the wires to the stick assembly will provide adequate protection against wire breakage.

Transmitter assembly

The transmitter RF deck has been assembled, aligned, and tested at the factory in accordance with FCC regulations. Don't tamper with any adjusting slugs on the RF deck — simply mount the deck as called for in the assembly manual. Component mounting, soldering, and interwiring techniques are exactly the same as those used in assembling the receiver-decoder. Again follow the assembly manual step-by-step.

The transmitter does have more mechanical assembly than the receiver-decoder. The antenna must be securely fastened to the case of the transmitter, the control sticks and their gimbaled mechanisms must be mounted on the face of the transmitter case, and meters and switches must be installed. Figure 5-7 shows typical assembly drawings for these mechanical sequences. All kit assembly manuals use exploded-view and other drawings liberally, so mechanical assembly is the fastest and easiest aspect of kit-building.

Servo assembly

To construct the servos, assemble the mechanical parts as shown in the assembly manual. Figure 5-8 shows a

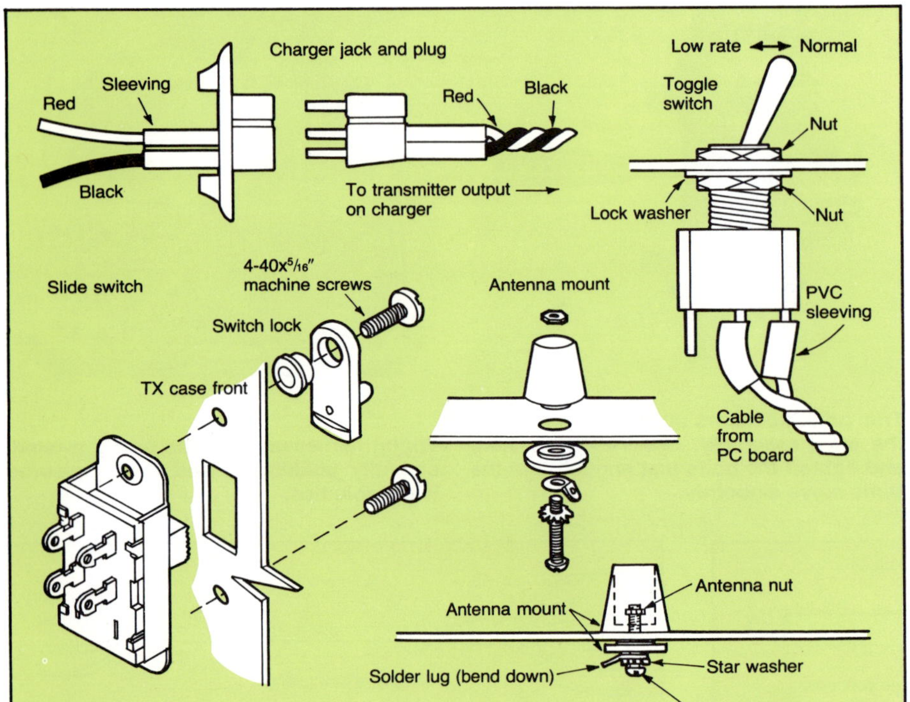

Fig. 5-7 TYPICAL MECHANICAL ASSEMBLY SEQUENCES

typical assembly drawing. Then build the servo amplifier PC board, employing the same techniques used in wiring the PC boards for the transmitter and receiver. Don't forget to mount any components that go beneath the IC before mounting the IC. Measure, tin, and solder all interconnecting wires. Clean the PC board with denatured alcohol and inspect for solder bridges, bad solder joints, and unsoldered leads.

Solder the pot leads to the proper pot terminals *as instructed for your kit.* Servos in the same line of kits may have different motor and pot connections, depending on which motors and pots were available to the manufacturer. Likewise, solder the motor leads in place. Finally, solder the twisted power and signal leads to their connectors and terminations.

Three-wire servos have a red, a black, and a third colored wire. Four-wire servos have a white wire in addition. MRC five-wire servos also have a green wire. In most systems, the red wire is +4.8 V, black is 0 V, white (if used) is +2.4 V, and the other colored wire is signal input. The only exception to this coding system that I know of is the RS Systems' servo — if you use this, follow the manufacturer's instructions. If any doubt remains, use a VOM to check for proper polarity and voltages.

For the Silver Seven the red wire from the decoder goes to +4.8 V, the black to 0 V, and the third wire goes to signal input. For other systems, wire the +2.4-V center tap from the battery if it's used.

If you're building a replacement servo for a system that had four servo wires, ignore all but the +4.8-V (red), 0-V (black), and signal (color varies) wires, because all currently manufactured kit servos need only these three wires. If you're using an MRC five-wire servo, connect the red and the green wires from the servo to the red wire from the receiver.

Please recall also that the Hobby Lobby sets don't use a servo amplifier in the servo; it's in the receiver.

Double-check all wiring before operating the system because reverse polarity can damage most servos.

Stop at this point. Final assembly of the servo takes place during system setup.

Battery pack assembly

Nickel-cadmium cells are wired in series, placed in plastic cases, and fitted with appropriate cables to make the transmitter and receiver battery packs. Begin assembly of 100-mAh and 225-mAh packs by soldering jumper wires to the solder tabs on one pair of cells. Hold the cells together with tape while soldering and maintain proper polarity — the shouldered end of the cell is positive, the flat end is negative.

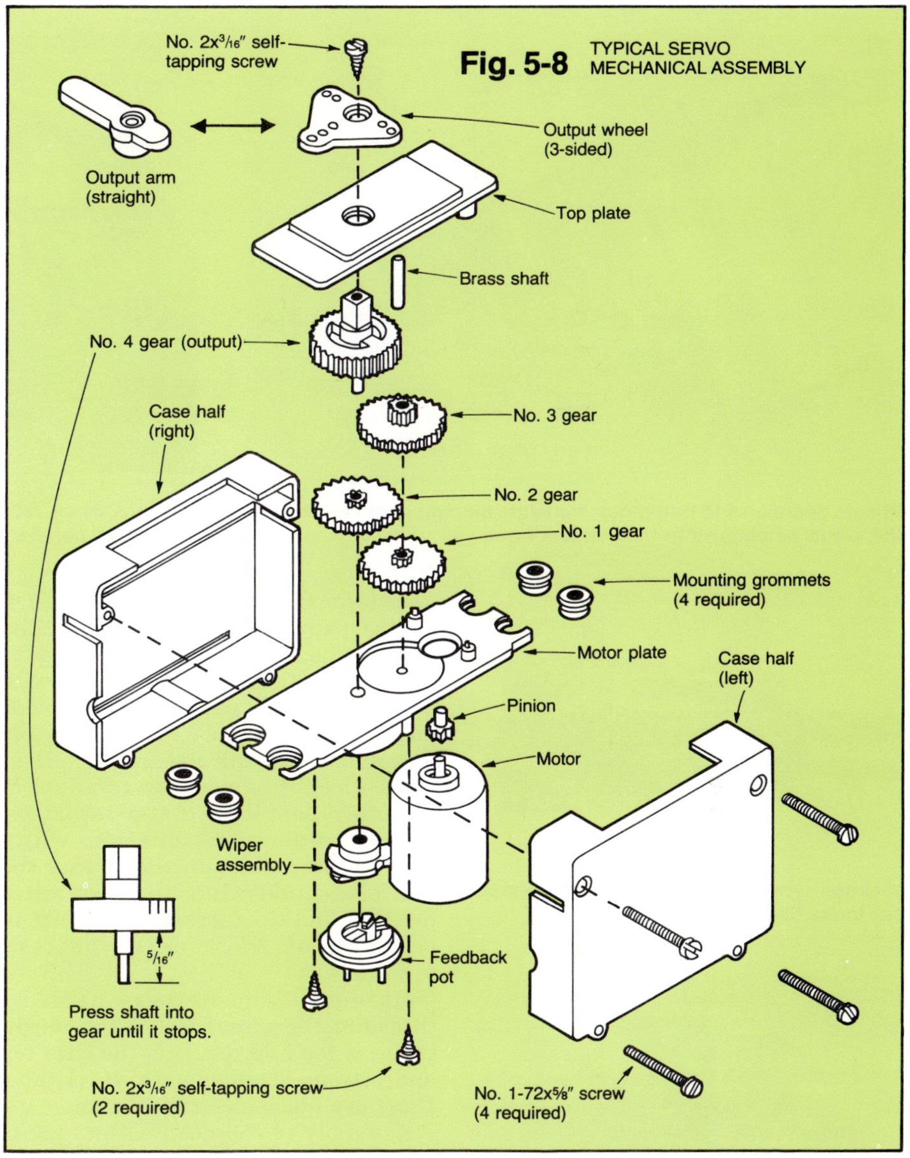

Fig. 5-8 TYPICAL SERVO MECHANICAL ASSEMBLY

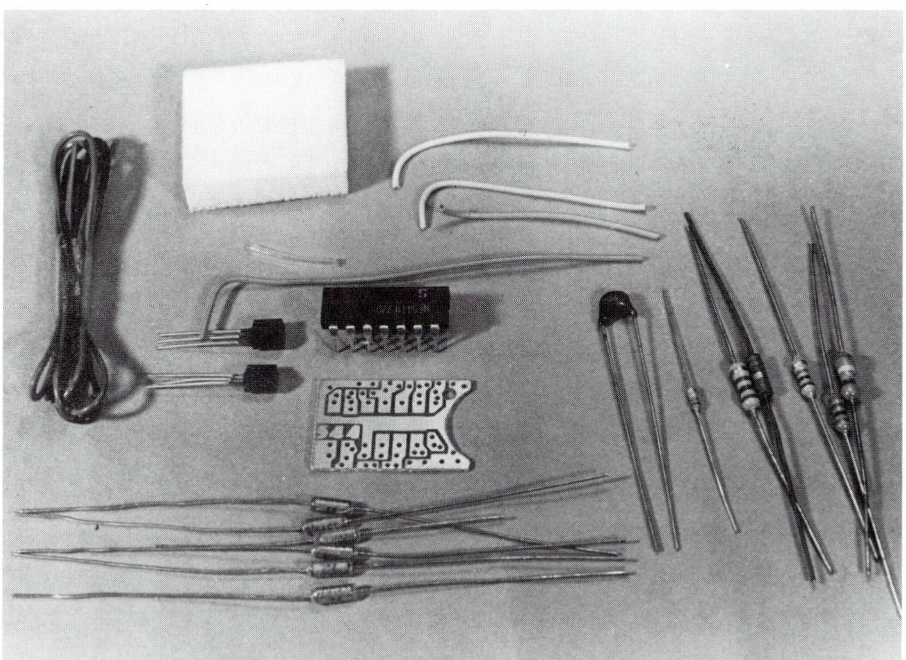

There aren't many components in a servo amplifier. Two external PNP transistors allow the servo to produce up to 7.5 pounds of output thrust.

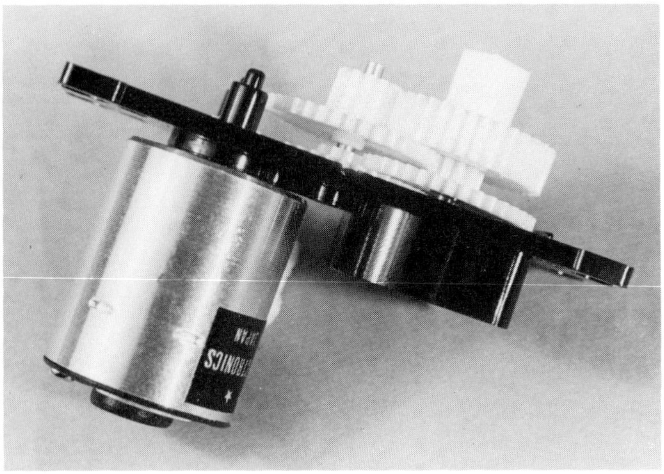

Plastic gears and 16 mm motor translate electric commands from the servo amplifier into mechanical force.

Hookup wires from the servo feedback potentiometer should always be parallel to one another.

A properly wired three-pin female connector looks like this.

Lacing around all the wires connected to this miniature switch prevents unwanted flexure that could lead to broken wiring.

The concave indentation on the PC board cuddles the motor housing.

The completely assembled servo amplifier shows how motor wiring is securely attached to PC board.

Repeat for the other pair of cells. Turn the two pairs of cells over and bend the solder tabs over upon themselves to prevent shorting. Install a jumper wire between the two pairs of cells. Solder an 8″ length of red wire to the positive terminal of one pair and an 8″ length of black wire to the negative terminal of the other pair. If a center tap is required, solder an 8″ length of white wire to the center junction. Twist the red, black, and white (if used) wires together and tie a strain relief knot in them about ½″ from the battery terminals.

Assemble the battery pack by threading the wires through the hole in the case top and securing the case top and bottom together with vinyl tape. Don't use black electrician's tape.

Assembly of 450-mAh battery packs is slightly different because of the pack's unusual configuration. Wire the cells together with jumper wires, being careful to maintain proper polarity. As with other cells, the shouldered end is positive, the flat end is negative. Use a piece of duct tape between the solder lugs and the cells for insulation. For strain relief, tie a knot in the red and black wires before they exit the hole in the case top, use a gob of RTV Silastic or silicone glue to fasten the wires at this hole, or tape the wires to the case after assembly. Secure the case together with vinyl tape, twist the leads together, and attach the female connectors. Assemble the charger, and

charge all new battery packs for 24 hours.

Transmitter adjustment

The system must be tuned before it can be used. The transmitter is tuned before the receiver because a functioning transmitter is required to align the receiver. Servos are adjusted last because we need both a functioning transmitter and receiver to test them. Although most kit-built systems are designed to be tuned with no more equipment than a good multimeter, an oscilloscope speeds the process.

Charge the transmitter and airborne battery packs as directed in the assembly manual, plug in the batteries, and you're ready to adjust and check out the transmitter. The Silver Seven's setup is typical of most kits. Turn on the transmitter and use a voltmeter to verify the presence of 9.6 V to the switch, RF deck, and encoder board. Verify V_R as 5 V $\pm$ 0.3 volt at pin 15 of the NE5044 and verify $V_R/2$ at the junction of the two 15-kΩ precision resistors as well as at pin 6 of the NE741 IC. This should be exactly half the voltage appearing at pin 15 of the NE5044.

To make final adjustments to the Silver Seven encoder, connect the ground lead of a vacuum-tube voltmeter (VTVM), a digital voltmeter (DVM), or a multimeter (VOM) with an impedance of at least 20 kΩ per volt to the junction of the precision resistors. Using a 1-volt or lower scale on the meter, set all centering pots, control pots, and trim pots so that their outputs are as close as possible to zero with respect to $V_R/2$. Set R10 and R11, adjacent to the NE5044, to their resistance center position.

Using an oscilloscope with an accurately calibrated time base, or the assembled and calibrated servo provided in the kit, make the following adjustments:

- Set the dual-rate switches to high rate. Set the high-rate pots fully clock-

A Mylar sheet separates the servo amp components and the feedback pot to prevent short circuits.

The high-strength plastic case bolts together to protect internal parts against dirt and vibration.

wise on all channels except channel 5 (retractable landing gear).

- Connect the scope or turn on the receiver and servos. Adjust R10 and R11 (they interact) to give a channel output of 1.5 ms at center and to produce $\pm$ 0.6 ms when measured at channel 1, the master channel. These settings correspond to mechanical center for the calibrated servo with travel of about $\pm$ 50 degrees. This excess travel is then eliminated to bring servo travel to the correct $\pm$ 45 degrees ($\pm$ 0.5 ms pulse width) by rotating the channel 1 range pot counterclockwise.
- Move the channel 1 reversing switch back and forth and observe that servo neutral is maintained. If not, adjust the elevator control pot tab until no motion is observed. Then readjust R10 and R11 slightly to regain correct center and throw.
- Move the calibrated servo from channel to channel and set each channel. Reset each control pot tab for channels 1 through 4 until all reversing switches operate without servo shift. As you do this, set the range for each channel for a nominal $\pm$ 45 degrees ($\pm$ 0.5 ms) throw.
- At channel 5, set one retract pot for +45 degrees rotation and the other retract pot for -45 degrees.
- If you want the auxiliary (AUX) channels to be precisely centered (versus deliberate offset), carefully set the AUX control tabs to center and adjust their centering pots in minute increments until their respective range pots can be rotated from end to end with no shift in servo position. Full tab movement can then be set to give the desired amount of travel, ranging from 0 to $\pm$ 60 degrees or more.

This completes the basic encoder setup. Low-rate on channels 1 through 3 is set by actuating the dual-rate switches and rotating the dual-rate range pots to give the desired setting. These may be readjusted, if desired, while the system is in use.

Receiver alignment

Prepare the receiver for tuning by running the tuning slugs up and down the tuning coils (L2 and L3/L4 in the Silver Seven) a few times with a small screwdriver. This establishes threads in the coil forms and makes for easier tuning. Clip one lead on each of two 10-μH chokes to ¼″. Solder them to the lands on the PC board as indicated in your kit's instructions. Put a piece of plastic tape between the body of the choke and the PC board to prevent shorts while tuning. If you don't already have a suitable nonmetallic tuning wand, whittle one from a wooden dowel or plastic rod. Use it for tuning both the IF transformers and the RF tuning coils. The receiver can be tuned most quickly and accurately using an oscilloscope. Connect the scope to the two chokes just installed. Set it to 0.1 V per cm or to the lowest setting available. Position the receiver on a pad of foam plastic on a nonmetallic workbench well away from objects such as pliers and soldering irons. Drape the antenna over the edge of the workbench. Collapse the transmitter antenna to its shortest length — usually 8″ to 10″. Turn the transmitter on. Apply 4.8 V to the receiver.

As you perform the following steps, increase the distance between the transmitter and receiver to maintain the displayed signal at about 1.5 cm peak-to-peak. After the first pass through the tuning process, remove the transmitter antenna to produce minimum radiation. If it is mounted internally, collapse it as far as possible. Do not touch the receiver or any of its leads while making adjustments. Using a nonmetallic tuning wand, adjust IF transformers T1, T2, and T3 for peak displayed output. Then adjust tuning coils L2 and L3/L4 for peak output. Keep moving the transmitter farther away and run through these adjustments at least three times to ensure that the receiver is peaked. The displayed signal must not be distorted and should have the waveforms shown in your kit instructions. Final tuning should be done with someone holding the transmitter — this simulates field conditions. After tuning, remove the 10-μH chokes. Finally, drip a little molten candle wax on each IF transformer slug to lock it in place.

If you don't have access to a scope, but do have a VTVM or VOM with a DC sensitivity of at least 20 kΩ per volt, set the meter to its lowest DC setting and tune the receiver as follows: Prepare the receiver and position the transmitter exactly as described above, but adjust the IF transformers and the tuning coils for the *lowest* voltage reading. Again, tune at least three times, and have someone hold the transmitter during the final tuning. After you've tuned the receiver, turn the transmitter off and observe the voltmeter. The voltage with no incoming signal must be stable. Your kit's tuning procedure may be a little different from the techniques explained here, so read the kit instructions carefully.

Servo adjustment

Final check-out and adjustment of the servos is made with the feedback pot mounted and positioned as directed in the instructions for your kit. This varies from type to type. Do not mount the motor yet. Arrange the partially assembled servo so that the PC board and wires do not short to the motor or to any metal objects on the workbench. Turn on the transmitter, then plug in the receiver. The servo motor should run one way or the other. Move the control for the channel selected back and forth. The servo motor will probably slow, stop, and then reverse at some point, hopefully near the center of stick travel. Set the control at neutral. Now slowly rotate the servo output arm (the gear train will move) until the motor

Four 1.2-V 450-mAh General Electric nickel-cadmium cells wired in series make up this 4.8-V receiver battery pack.

stops running. The output arm should be near center. If it is not, position the output exactly at the designated center position. Hold it in that position, and use a small jeweler's screwdriver (most servos) or a small hex wrench (Dunham servos) to adjust the pot until the motor stops. Turn off the receiver. Position the motor and press it into place in the servo case. Be sure the motor pinion meshes smoothly with the first larger gear. Turn on the receiver. If all is well, the servo will center and sit quietly. If it runs to one end and refuses to budge, you've wired the motor or pot connections backward. Turn off the receiver and reverse one set of wires, but not both. Remove the motor and gently center the servo output arm. Reinstall the motor and turn on the receiver.

Repeat this setup procedure for all servos.

The system is now ready to go. Save all assembly manuals because they contain valuable schematics and troubleshooting information that will help you identify the cause of most malfunctions. If you ever find that you can't identify the cause of a problem, don't tinker with the system and *hope*. Send it for service as described in Chapter 1.

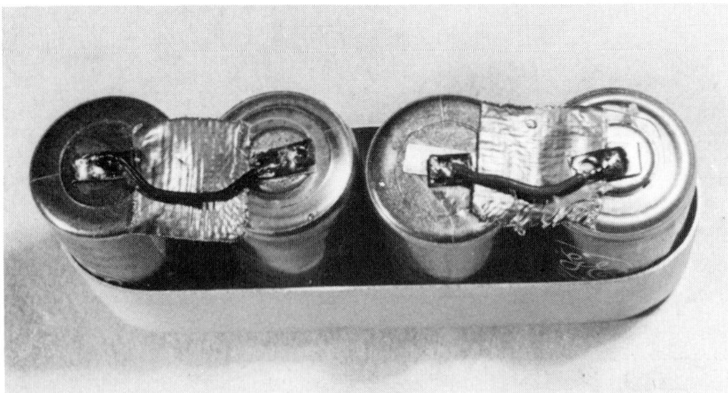

Connect the cells with stranded hookup wire and use heat-resistant duct tape for insulation. Don't solder to the body of the cells, use the solder tabs built into each cell. Hold the cells together with vinyl tape.

Tie a strain-relief knot in the wires just before they exit the top of the case.

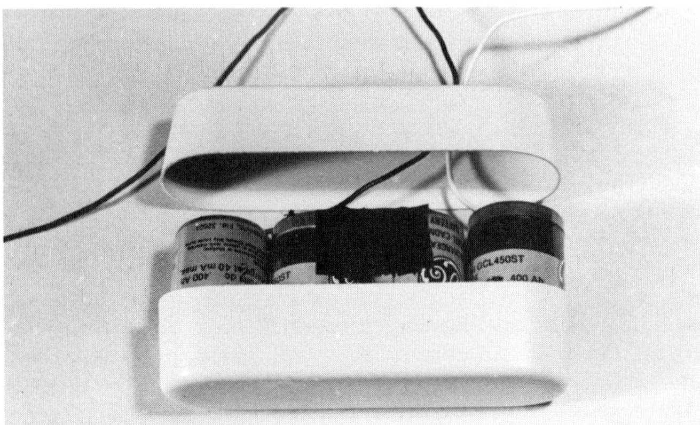

Two four-cell assemblies make up the 9.6-V transmitter battery pack. The white wire and the free wire will go to the second assembly to complete the series connection.

Tape the two assemblies together, wire to the power switch, and install the pack in the transmitter.

6 Maintenance

Integrated circuits have increased the reliability of RC systems by a factor of 1000, so crashes caused by failure of the RC gear are now a minor percentage of all crashes. Most crashes today are caused by pilot error or structural failure in the model. Nevertheless, because no man-made system is 100 percent reliable, because our model RC systems seldom have redundant features for fail-safe operation (we can't afford the cost or weight), and because failure of the RC system can have catastrophic results, it's only reasonable that we try to prevent as many RC system failures as possible.

Reliability begins with good design, sound manufacturing techniques, and careful post-assembly testing to detect early failures. Contemporary RC systems are well-designed and carefully manufactured, so aside from an occasional defect that slips by a factory's quality control team, this need cause little concern. You become responsible for reliable operation after taking delivery of the system. You'll have little fear of catastrophic system failure if you:

- Take care not to expose the system to physical environments it wasn't designed to survive.
- Burn in the new system before installing it in a model.
- Establish checklists to verify proper operation of all subsystems before each flight.
- Routinely inspect and overhaul those parts known to be least reliable or most subject to wear.

Care and burn-in

The first of these points is simplest: Don't abuse the system. RC systems are not designed to be dropped on hard floors, or to be subjected to temperatures below 0° F or above 120° F to 150° F. Your system's reliability increases dramatically if you don't store it inside a car trunk on a summer day or leave it in an unheated garage in the winter. Batteries are not designed to survive complete lack of attention for months on end. They will last far longer if given the routine maintenance described later in this chapter.

The second point, burn-in, is also simple. After taking delivery of a new system or after assembling a kit-built system, exercise the servos several hundred times, checking especially carefully for irregular motion. Turn the transmitter and receiver on and off 100 times or so. See how long the transmitter and airborne battery packs operate before falling below the voltage necessary for safe operation. Nickel-cadmium batteries produce a nearly level output until shortly before exhaustion, after which output drops off steeply, so watch for this "knee."

Checklists

The third point, checklists, is crucial.

- Before leaving home, be certain that the battery charger has charged the batteries. If the charger has a light or meter, does the light or meter show that it is working?

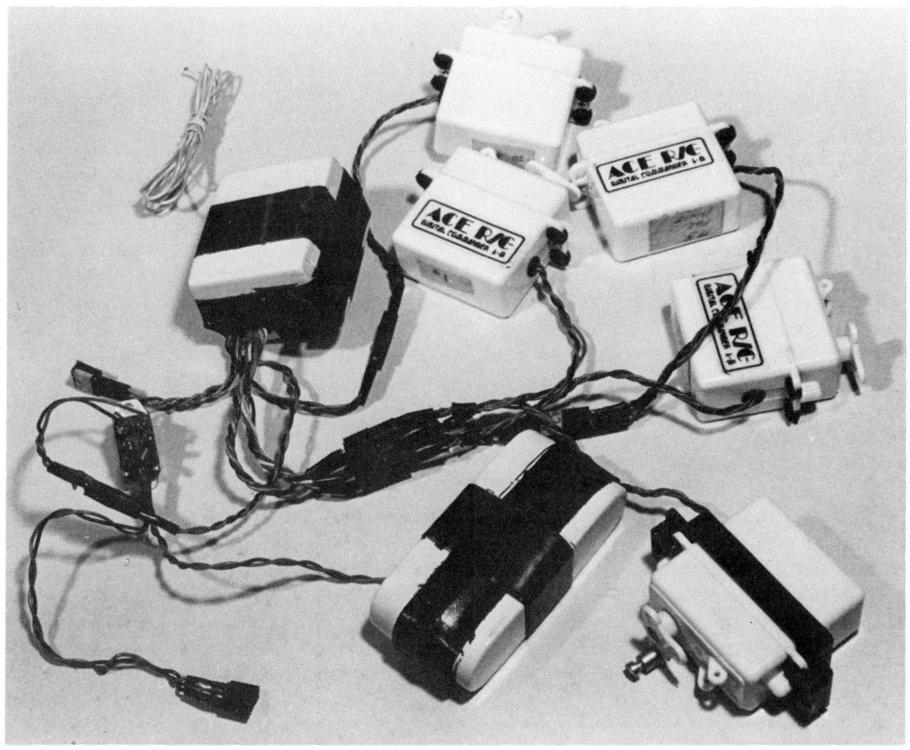

The airborne portion of an RC system needs frequent inspection. Look for broken wires and loose connections.

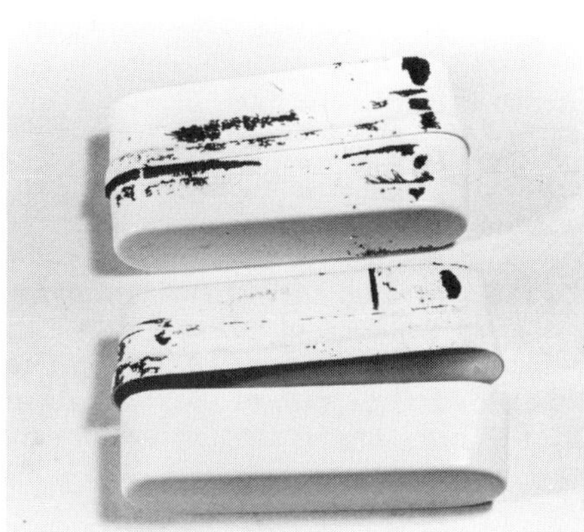

If you choose to replace only the bad cell in a battery pack, disassemble and clean the case before proceeding further. This case was hard to clean because it was held together with black electrician's tape. Use 3M vinyl tape instead.

Begin maintenance on a servo by removing the case. Put the bolts in a jar so they won't be attracted to the motor magnet.

Carefully disengage the motor and spread out all parts for close visual inspection with a jeweler's loupe or other strong magnifier.

- At the field, see if the transmitter operates properly. Turn it on, with the antenna collapsed, and watch the output meter or ESV for a few minutes to be certain output remains constant. Occasionally, one cell in a battery pack goes bad during charging. If it has, you want to find out now.
- Leave the transmitter on, and turn on the receiver. Move each transmitter control smoothly and slowly over its full range and check that each servo responds smoothly. Watch also for jumping motions in servos not being commanded. Such movement often indicates a noisy battery or a servo motor or feedback pot that is beginning to fail.
- With the transmitter and receiver still on, and the transmitter antenna fully collapsed or removed, range-check the system. Always perform range checks in the same location. Verify that no RF interference is present. If the collapsed antenna range has decreased by a third or more since the last range check, test again with the transmitter antenna extended. You should have full control for at least 500 feet.
- Inspect the receiver antenna for kinks, frays, and breaks. Inspect all visible connectors and wires. Inspect all control surface linkages.
- If the model has ailerons, plug in the aileron servo, check that the connector is tight, and, after attaching the wing, check for correct motion. Be certain the aileron connector cable isn't pinched between the wing and saddle. You say the ailerons move when you move the retract switch? Well, you somehow got hold of the wrong connector. You should have checked that *before* you put 52 rubber bands on the wing!
- Start the engine and recheck all servo movements at several throttle settings. If you suspect a vibration problem, have a helper hold the model while you repeat the range check with the motor running.

These checks should be extended, to some degree, to each flight during an operating session. At the least, don't leave the pit area until you have turned on the transmitter and have observed the meter for at least 15 to 30 seconds, turned on the receiver (yes, many gliders have gone on their way with the receiver turned off), and checked the movement of all control surfaces and the throttle. If you operate for more than an hour, check battery packs with an ESV between flights. If you use 100-mAh or 225-mAh airborne battery packs, consider recharging them between flights. I've used an airplane model as an example, but these checks should be performed for cars, boats, and other types of models as well.

Routine maintenance

The fourth point, routine maintenance, is important enough to be broken down into subcategories: care of the batteries, care of the servos, care of the receiver, and care of the transmitter. On the basis of information I've collected over the last 13 years, I've concluded that out of 1000 system failures, 747 are in the batteries, servos, and receiver, that is, in the subsystems carried in the model, Fig. 6-1. Of these, 438 are battery failures, 212 are servo failures, and 97 are receiver failures. Transmitter defects account for the remaining 253 failures. When the data are arranged in rank order, you can see that 50 percent of the problems are with batteries, that 5 to 8 percent are with servo feedback pots and servo motors, and that other components are much less likely to cause trouble. A plot of relative failures by component, Fig. 6-2, is even more revealing: Nine items contribute two-thirds of all failures, and just 12 items contribute almost three-fourths of all failures.

Batteries

Nickel-cadmium batteries have several virtues that make them useful for RC systems: They are lightweight in relation to storage capacity, are physically strong and vibration-resistant, are available in several convenient sizes, can be recharged in 14 to 16 hours (less for special designs), and are reasonably inexpensive. Their vices include an unpredictable time-to-failure, occasional refusal to charge completely, and a tendency to develop internal shorts and opens. They occasionally become electrically noisy. While nickel-cadmium batteries are far from perfect, they are the best we have. Vigilance is the secret to happiness with nickel-cadmium batteries. They should be regularly cycled, that is, charged and discharged under controlled conditions, in accordance with the manufacturer's instructions, and they should be frequently checked for loss of capacity and other defects. Battery cycling is most easily performed with a battery cycler such as the Ace Digipace I. This device charges nickel-cadmium batteries at any of several charge rates, then automatically discharges them through a controlled resistance. In this way you can determine discharge time and battery capacity. A new 450-mAh battery pack should have a discharge time of about 100 minutes before the voltage drops to the

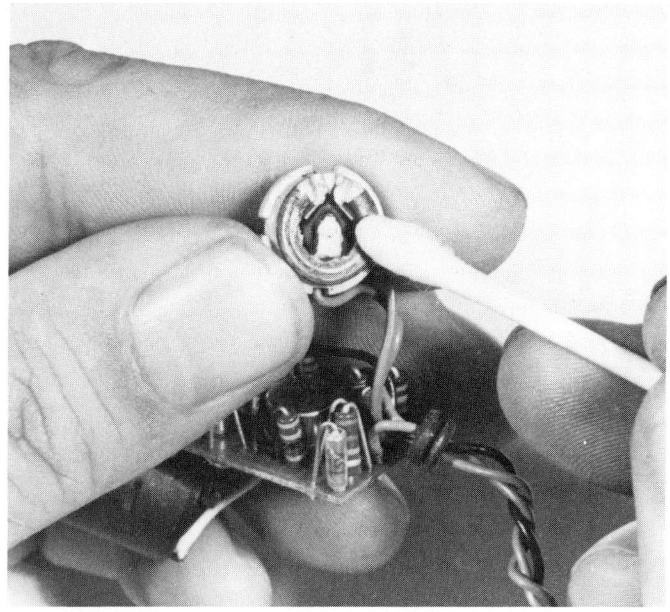

(Left) Mark the position of the pot element for future realignment. (Above) Loosen the screws that hold it in place and remove the potentiometer.

Clean the pot element with a Q-tip wetted with dope thinner or some other degreasing solvent.

Place a tiny amount of silicone-based lubricant such as Outers gun oil on the pot element, including the center contact. Wipe off any excess lubricant.

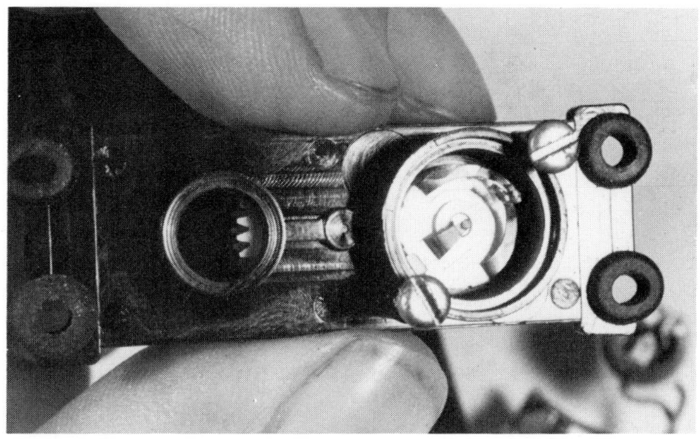

This brass wiper is only moderately worn, but I'd replace it anyway just to be safe.

end-of-capacity level of 1.1 V per cell. If the discharge level ever decreases to less than 75 minutes, the pack has gone bad and should be replaced. Replace the entire pack unless it has been used for less than 6 months. If the pack is less than 6 months old, place it under load with the cycler, and check the voltage of each cell. After a full charge and 15 minutes of discharge on the cycler, individual cells must produce at least 1.2 V. If a cell fails the test, replace it with a new cell of the same make and capacity — these are usually available at your hobby shop or from the manufacturer. If none is available, Lafayette and Radio Shack carry replacement cells that will serve in an emergency. After a cell is replaced, cycle the entire pack to check capacity.

If you don't have a battery cycler, you can still verify capacity reasonably well. Ohm's law states that current flow equals voltage divided by resistance, or, $I = \frac{E}{R}$, where I = amperes, E = volts, and R = ohms. Most cyclers discharge the battery on a load of 200 milliamperes, so use that load for manual discharging. For a 9.6-V transmitter pack, $0.2\text{ A} = \frac{9.6\text{ V}}{R}$, so that $R = 48\ \Omega$. There is no 48-ohm fixed resistor, but 47 ohms is a standard value and is close enough. The resistor must be large enough to dissipate the power absorbed as heat during discharge. Its wattage must be $W = I^2R$, so $(0.2)^2(48) = 1.92$ W. Use a 3-watt resistor for safety's sake. In this example, a 47-ohm, 3-watt resistor with a 5 percent tolerance serves nicely.

In the same way, the 4.8-V receiver pack should be discharged through a 22-ohm (also a standard value), 2-watt, 5-percent resistor. These formulas can be used to determine the resistance and wattage of resistors used to discharge batteries of other capacities. In all cases, discharge the battery through the resistor, checking that the voltage stays at or about 1.2 V for at least 75 minutes.

During normal operation, when you're using the system one or more times a week, the batteries should be charged and tested before and after each operating session. If more than two days pass between sessions, charge the batteries before flying. During off seasons, cycle the batteries once a month to ensure they don't go below 1.1 V per cell. This off-season monthly charging is necessary because nickel-cadmium batteries self-discharge at a rate of approximately 1 percent of capacity per day at room temperature, more rapidly at higher temperatures. If battery packs do drop below 1.1 V per cell, as might happen if you accidentally leave the receiver on for several hours, or if you forget the pack's once-a-month off-season charge, cycle them at least once before use.

Shorted and open cells give abnormal readings on a voltmeter or ESV, so are easily detected. Unfortunately,

Group°	Rank	Component	Failures in 1000
I	1	Loss of receiver battery capacity	251
IV	2	Loss of transmitter battery capacity	160
III	3	Dirty servo feedback potentiometer	46
III	4	Worn or dirty servo motor	42
I	5	Broken receiver battery pack wire	41
I	6	Poor contact in receiver power switch	36
I	7	Open cell in receiver battery pack	31
II	8	Broken power, servo, or antenna wiring in receiver	29
I	9	Oil in receiver switch contacts	29
III	10	Impact damage to servo gear train	26
II	11	Semiconductor failure in receiver	25
I	12	Shorted cell in receiver battery pack	20
III	13	Failed solder joint in servo	15
III	14	Worn servo feedback potentiometer element	15
III	15	Worn servo feedback potentiometer wiper	15
II	16	Failed solder joint in receiver electronics	15
V	17	Failed transmitter RF power amplifier	12
V	18	Poor connection or broken transmitter antenna wire	11
I	19	Poor connection to receiver power terminals	10
I	20	Dirty receiver switch contacts	10
III	21	Worn servo gear train caused by vibration	10

Fig. 6-1 FAULT TREE OF SYSTEM FAILURES

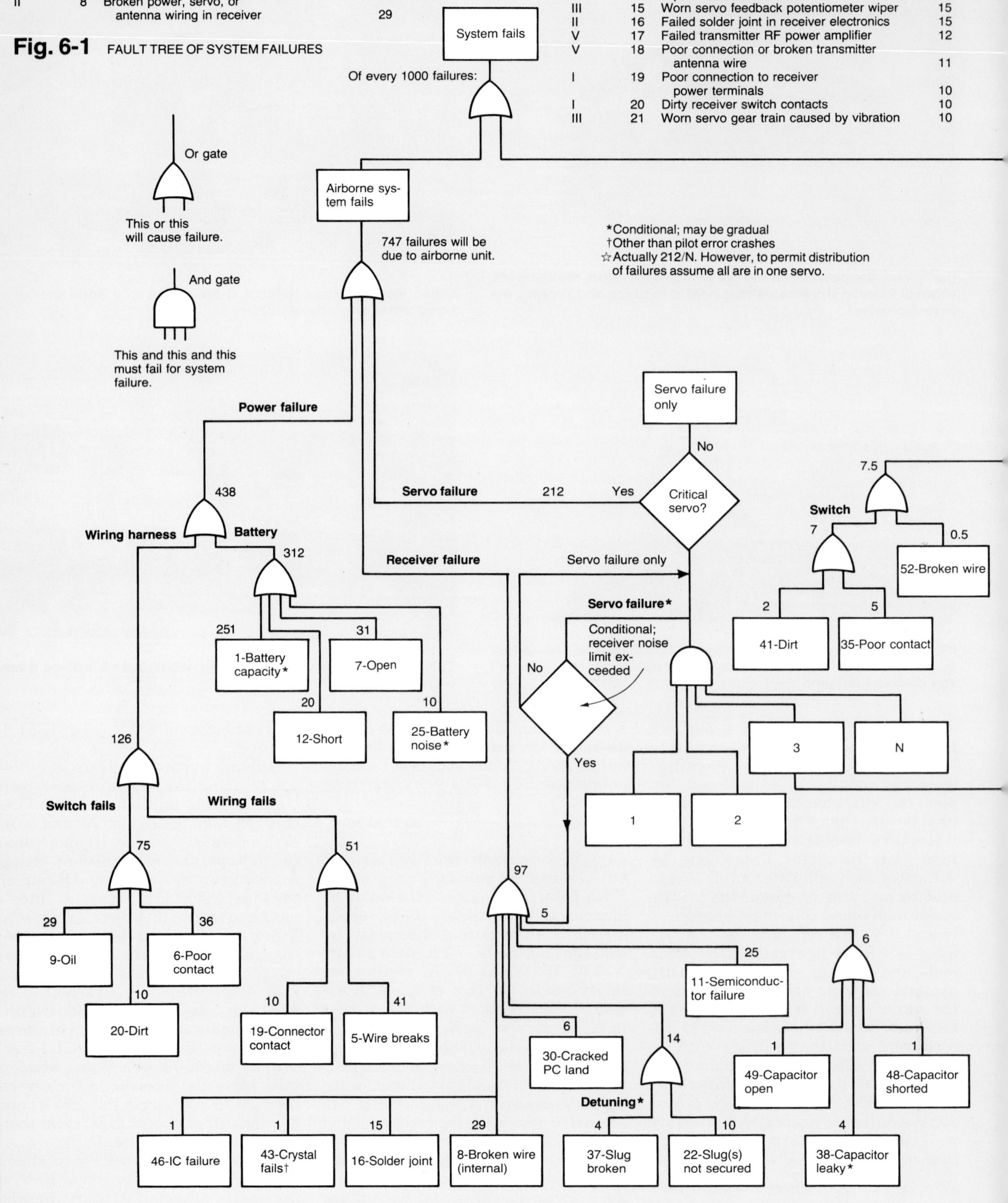

Group°	Rank	Component	Failures in 1000
II	22	Loose receiver IF transformer tuning slugs	10
III	23	Failed servo IC	10
IV	24	Shorted cell in transmitter battery pack	10
I	25	Noisy cell in receiver battery pack	10
VI	26	Failed semiconductor in encoder	10
IV	27	Noisy cell in transmitter battery pack	10
III	28	Open in servo motor commutator	9
VI	29	Broken wiring in encoder	7
II	30	Broken land on receiver PC board	6
III	31	Broken servo feedback potentiometer wire	5
III	32	Loose servo wiper holder	5
III	33	Broken wire in servo electronics	5
III	34	Bad capacitor in servo electronics	5
IV	35	Poor contact in transmitter switch	5
IV	36	Open cell in transmitter battery pack	5
II	37	Broken receiver IF transformer tuning slugs	4
II	38	Leaky capacitor in receiver	4
III	39	Broken component in servo electronics	4
VI	40	Worn potentiometer in encoder	3
IV	41	Dirty transmitter power switch	2
VI	42	Leaky capacitor in encoder	2
II	43	Failed receiver crystal	1
VI	44	Failed IC in encoder	1
V	45	Failed transmitter RF section component (not amp)	1
II	46	Failed IC in receiver	1
VI	47	Broken component in encoder	1
II	48	Shorted capacitor in receiver	1
II	49	Open capacitor in receiver	1
VI	50	Shorted capacitor in encoder	0.5
VI	51	Open capacitor in encoder	0.5
IV	52	Broken transmitter power wiring	0.5
V	53	Failed transmitter crystal	0.5

°Group I = Receiver power supply
Group II = Receiver
Group III = Servo
Group IV = Transmitter power supply
Group V = Transmitter RF section
Group VI = Encoder

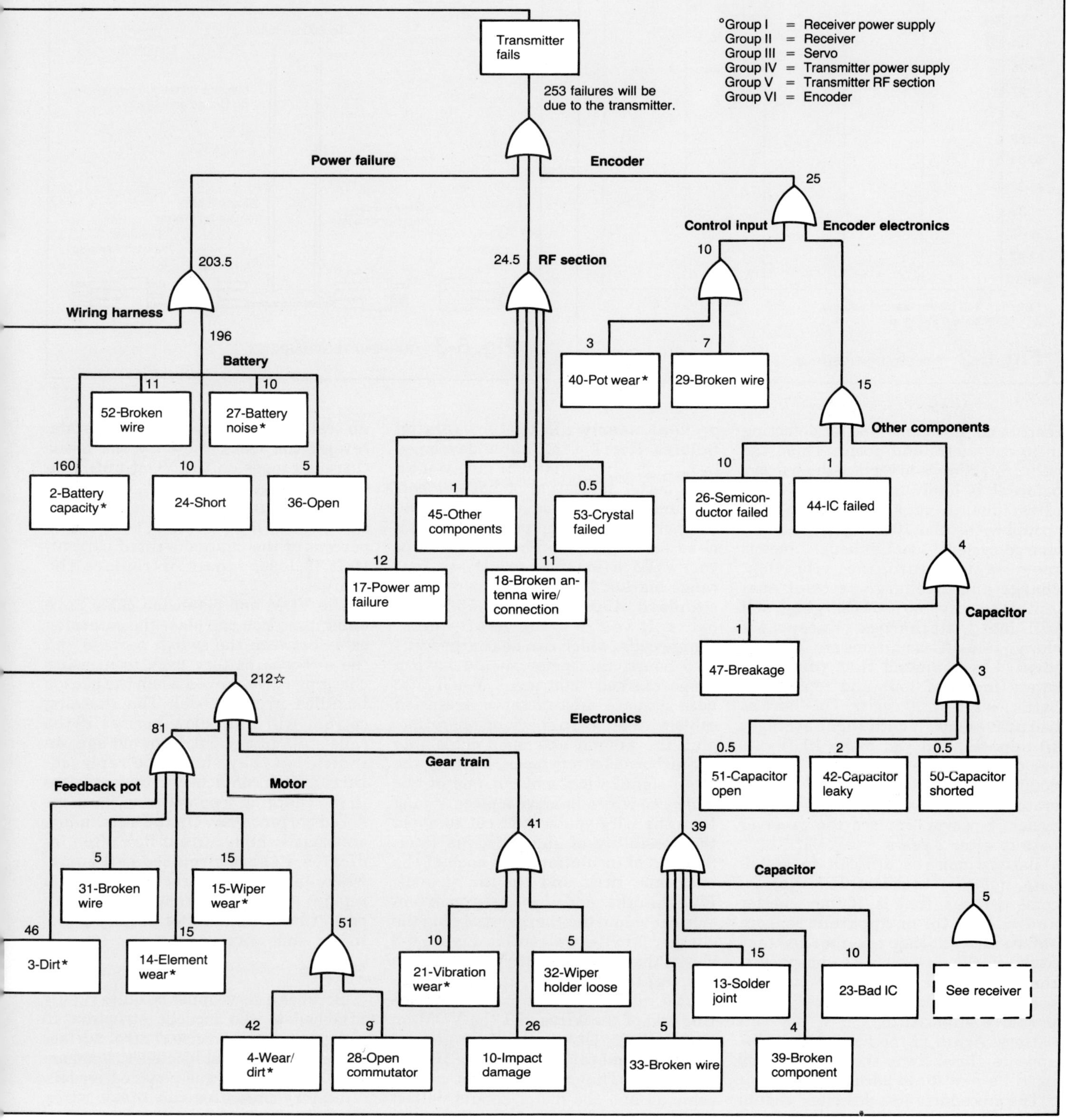

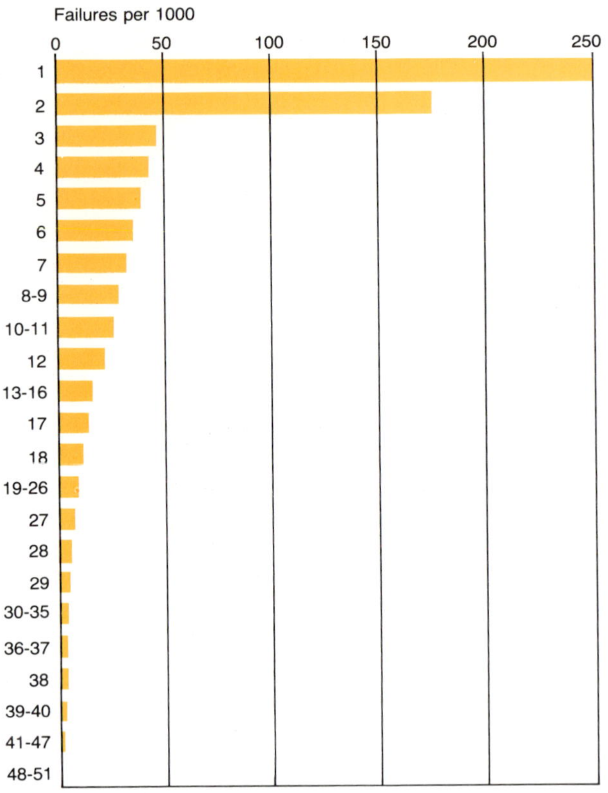

Failures listed by number correspond with table shown in Fig. 6-1.

Fig. 6-2 FAILURE DISTRIBUTION

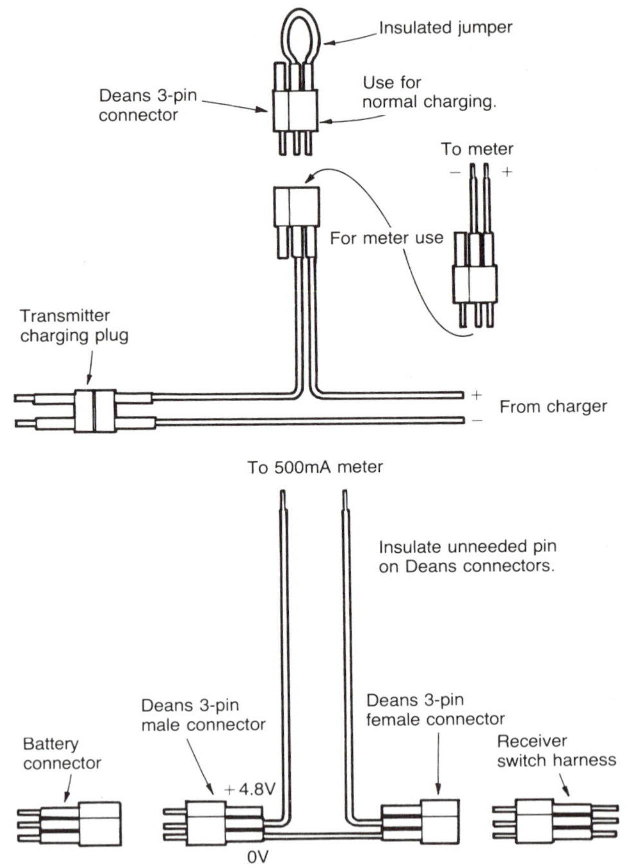

Fig. 6-3 CURRENT TEST HARNESS

there is no proven method for detecting incipient shorts and opens. Thus, the only protection is to replace the battery before it is likely to fail. In ideal applications, a nickel-cadmium cell should withstand 1000 charge and discharge cycles. "Ideal," though, means room temperature, no vibration, charge and discharge rates of one-tenth capacity, no overcharging, and full-depth discharges. Except for charge rate, RC conditions are far from ideal. I recommend that you use a safety factor of four and replace all batteries after 250 cycles. This isn't as bad as it seems. If each flight averages 10 minutes, and you make 10 flights per cycle, the batteries will last for 2500 flights. That is 5 years if you operate once a week. If you don't use a cycler, however, replace the receiver battery after 2 years.

Battery noise is difficult to anticipate, but easy to detect. If servos become nervous, that is, if they chatter and vibrate for no apparent reason, substitute another receiver battery pack. If the nervousness disappears, throw away the bad pack. If you suspect a noisy transmitter battery, temporarily substitute a 9-V dry-cell battery. Again, if the nervousness disappears, throw away the bad pack and install a new nicad pack.

The procedures just described should prevent nearly all battery-related failures. Here's a technique developed by George Myers that will give you an even better indication of battery performance and let you check that the charger is operating properly. You'll need an ammeter: The ammeter in your VOM is fine if it has a DC-current range marked "60 mA" if you're using standard 450-mAh and 550-mAh packs. If you're using Kraft quick-charge cells, which can be charged in 4 to 6 hours, the meter should have a range marked "300 mA." You'll also need a spare aileron servo extension cable with plug. To check on operation, plug the aileron extension cable into the airborne battery pack. Identify the servo signal wire, and cut one of the *other* two wires. Remove a piece 1″ long from the wire you've just cut to avoid the possibility of short circuits later. Strip ¼″ of insulation from each of the ends, and twist and tin the strands. Plug the other end of the aileron extension cable into the charger and plug the charger into the wall outlet. Figure 6-3 shows the arrangement. The charger's red light should not light, because you've interrupted the circuit by cutting one of the wires. Set the VOM on the highest DC-current range and touch the meter leads to the tinned wire ends. The charger's light should come on and the meter should deflect up-scale or down-scale. If down-scale, reverse the leads. Now try the lower current ranges on the VOM until you obtain the greatest meter deflection. Read the current indicated. Everything's okay if the reading is 90 to 100 percent of the charger's rated current. If it is less, repair or replace the charger.

The VOM and extension cable have other uses: You can place the extension cable between the switch harness and the airborne battery pack to measure charging current even when the pack is installed in the model. The charging current will fall below normal if the cells start to dry out from old age, an indication they should be replaced. Dirty switch contacts and broken wires in the charging loop will also cause reduced current flow. On the other hand, abnormally high current flow often indicates a short-circuited cell somewhere in the pack. Track it down and replace it, or simply replace the entire pack. Check transmitter battery packs in the same way.

Servos

Because a servo must be quite rigidly attached to the model's structure in order to ensure precise control surface movement, it is subjected to tremendous vibration in gas-powered models. Vibratory pressures can break wires,

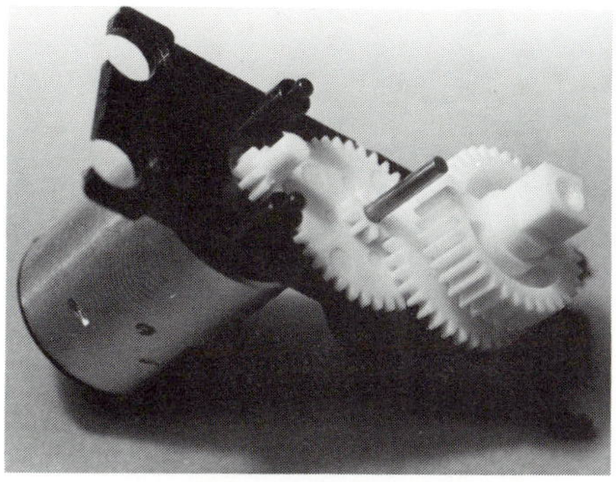

Disassemble the gear train, then clean and inspect each gear.

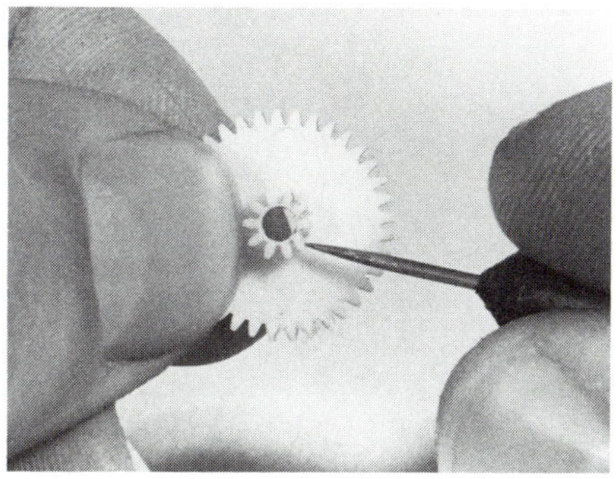

A broken tooth shows that this gear must be replaced.

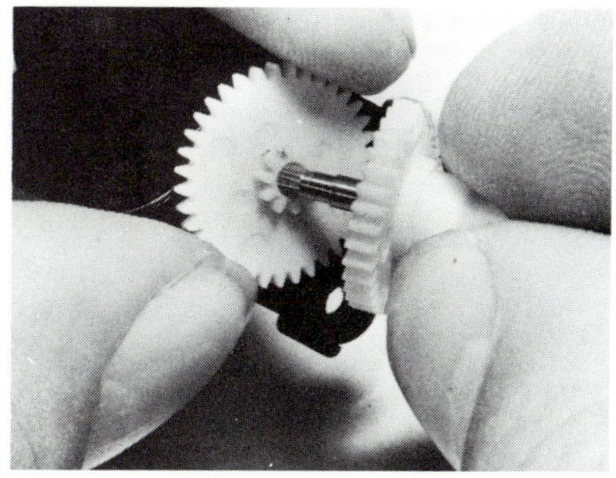

(Top) Be gentle when reinserting the pot wiper shaft and gears into the frame. (Middle) Each gear is different and has its own number. Be sure they are in correct order. (Bottom) Check the reassembled gear train for even gear meshing and smooth movement in both directions.

shake loose feedback pot wipers, damage gears, and generally wreak havoc with the servo. Therefore, servos should be inspected and overhauled after 20 to 30 hours of operation or 200 flights. I recommend that you purchase one or two spare servos and use these as temporary replacements while performing routine maintenance on the system's regular servos. Follow these steps:

Step 1. Discard the worn servo mounting grommets and disassemble the case by unscrewing the screws or bolts that hold it together. Tag all screws and bolts and place them in a small container so that they won't stick to the motor magnet when the motor is removed.

Step 2. Remove the motor and the feedback pot from the servo case. Do not unsolder the leads from the motor and feedback pot to the servo amplifier PC board. The motor is usually lightly press-fitted into the servo case — gently pry it out with a jeweler's screwdriver. The feedback pot is almost always held in place by two small sheet metal screws or 2-56 bolts.

Step 3. Clean the case and gears with rags and Q-tips wetted by trichloroethylene, acetone, dope thinner, Varsol, or denatured alcohol. Test first to be certain that the cleaning fluid does not dissolve the plastic parts. Scrub the solder side of the PC board with an old toothbrush wetted with the cleaning fluid.

Step 4. Working under a bright light, examine each tooth of each gear for roughness or excessive wear. Use a jeweler's loupe or magnifying glass. Control surface flutter and engine vibration cause heavy wear to the output gear at its neutral position, so examine that gear especially carefully. Inspect the feedback pot wiper and the wiper carrier to be sure they are properly seated on the output shaft that transmits motion from the output gear to the wiper carrier. Wiggle the output gear and shaft to check for excessive play. If the wiper holder wobbles noticeably, buy a new set of servo mechanics. Similarly, check each gear set for excessive wobble on idler shafts. Check that the idler shafts (usually small brass rods) are not bent and that the recesses for them in the servo case are not cracked or broken. Inspect the case and mounting lugs for cracks.

Step 5. Clean the feedback pot element with a Q-tip and one of the solvents mentioned above. Inspect the pot element, again using a jeweler's loupe or magnifying glass. Look for excessive wear around the neutral contact points, roughness anywhere around the track, cracks in the element, breaks or looseness in center and end termination posts, and ground-off brass particles from the wiper. Turn over the pot and check solder joints for crystallization or cracking. Look for frayed wire at the junction of solder and insulation. Resolder any bad joints and replace any frayed wire. Excess element wear, brass particles that won't scrub off, or cracked parts mean that the pot element must be replaced. The wiper, wiper holder, and pot element are sold only in sets and must be exact replacements. Adjust the wiper as shown in Fig. 6-4.

Step 6. Inspect the servo amplifier PC board for cracks, crystallized or broken solder joints, and broken components. Resolder suspicious joints and replace broken components. Replace any frayed wires. Replace any PVC sleeving that has hardened, using the same color sleeving to preserve the color coding system.

Step 7. Check that the servo motor current consumption is within normal values. You may use the extension cable and ammeter technique described above, or, for more accurate measurements, you may remove the motor from the servo amplifier by unsoldering the wires from the PC board to the motor. Sketch the connections on a piece of paper so that you'll get the wires back to their proper locations later. Connect the power supply or battery pack leads in series with the motor and VOM. With 4.8 V applied, the motor should draw 80 to 100 mA when running free. Gradually slow the motor by applying thumb pressure to the motor pinion. With slight pressure simulating normal servo operation, the current drain

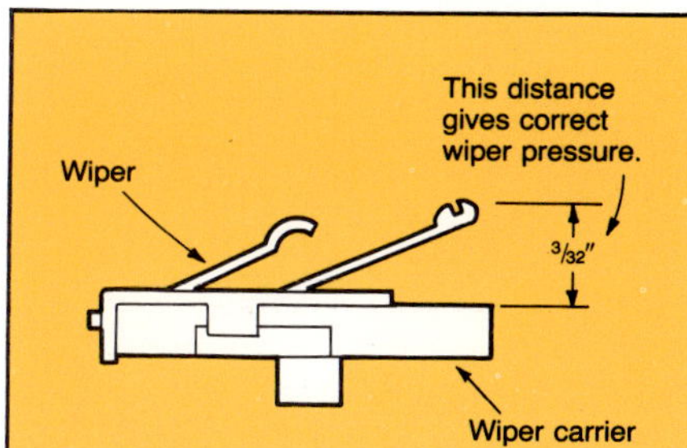

Fig. 6-4 WIPER PRESSURE

should reach 150 to 250 mA. As pressure is increased until the motor stalls, current drain should increase to 450 to 500 mA for motors with 8- to 12-ohms resistance and to nearly 600 mA for 6-ohm motors. A 25 percent variation from these readings indicates that the motor should be overhauled or replaced, as does a decrease of more than

Foam tape makes a good insulator between pot leads and servo amplifier PC board.

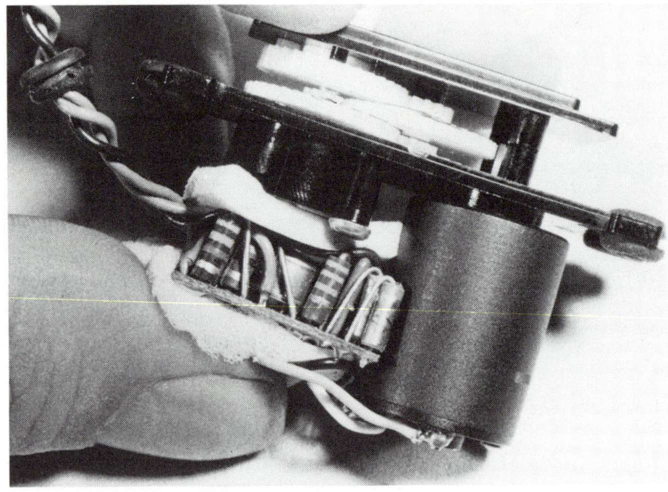

All broken parts have been replaced, all parts have been cleaned, and the pot element has been lubricated, so this servo is ready to go back into its case and return to operation.

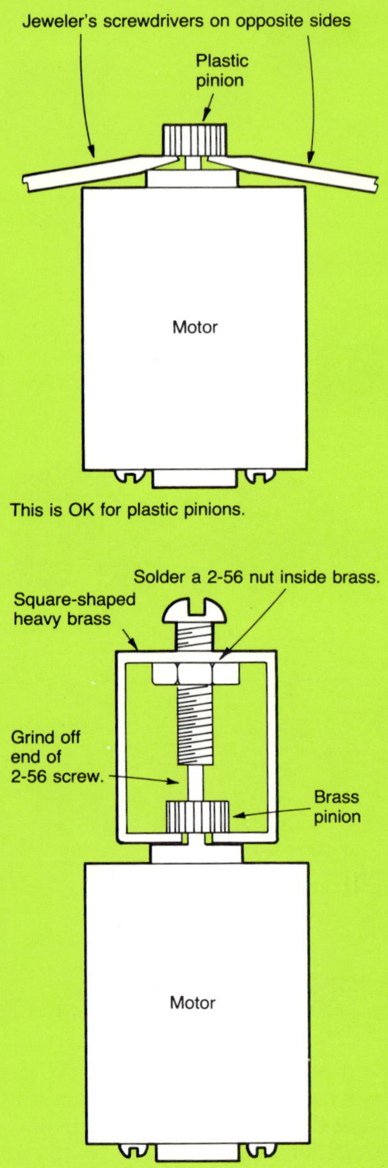

Fig. 6-5 GEAR PULLER

20 percent from the motor's speed when new. It is easier to replace than overhaul a worn motor; use an exact replacement or at least be certain that the new motor has the same diameter and resistance.

If you choose to overhaul the motor, get ready for a painstaking series of tasks. Ignore the advice sometimes given in modeling magazines about cleaning motors by dipping them into a nonflammable solvent and running them in the solvent to loosen grime. This shortcut will ruin the motor by drawing lubricating fluid from the Oilite motor bushings. Further, it won't remove the dirt, but will redistribute it inside the motor casing.

Begin motor overhaul by disassembling the motor. The removable end plate is always at the brush end and carries the brushes. The other end is a can-shaped motor housing. The armature, brush assembly, and end plate must be withdrawn from the housing. This requires that the pinion be removed from the armature shaft. A tiny homemade gear puller, Fig. 6-5, is best for this, although a plastic pinion can be removed by gently prying it off with two jeweler's screwdrivers used as levers on opposite sides of the pinion. Brass pinions are harder to remove, so use the gear puller whenever possible.

Remove the screws that hold the end plate in place. Mark the end plate and can-shaped motor housing positions relative to one another for realignment during reassembly. Place the motor end plate on a foam plastic pad, hold the end plate/armature/shaft assembly on the pad with a T-pin, straightened paper clip, or small hex wrench, and carefully lift the motor housing clear of the core assembly. Remove the end plate from the armature assembly by lifting the brushes clear, simultaneously sliding the end plate off the shaft. Clean the commutator, brushes, and motor housing with Q-tips and one of the solvents mentioned earlier. Wipe the bushing to remove dirt and excess oil. Chuck the armature shaft in a variable-speed motor tool, cut a 6″ strip of No. 600 emery cloth or wet-or-dry sandpaper, rotate the armature at the motor tool's lowest speed, and gently burnish the commutator. If you don't have a variable-speed motor tool, simply rub the emery cloth or sandpaper over the commutator in a sawing motion. Clean the burnished commutator

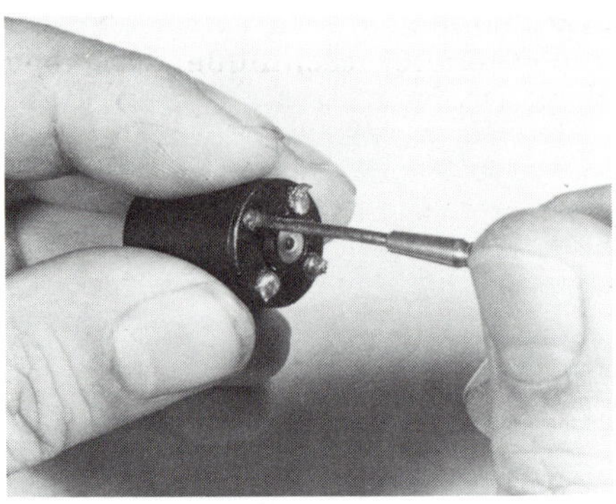

Use jeweler's screwdrivers to remove the tiny screws that hold the motor end plate to the motor housing.

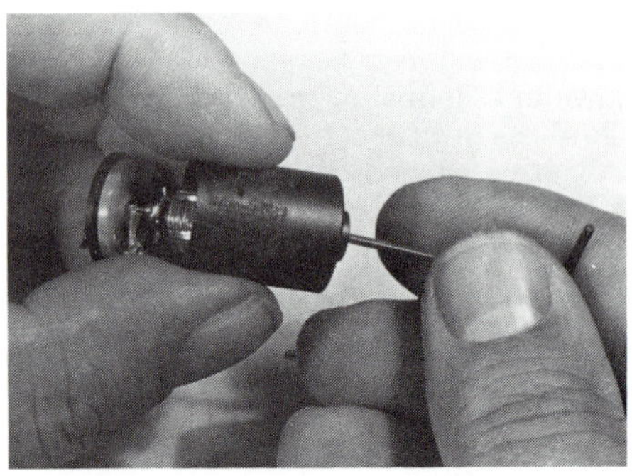

The plastic pinion on this Dunham servo was easy to remove with a jeweler's screwdriver. I used a 1/32″ hex wrench to push the shaft through the hole.

with Q-tips and solvent. Then flush it with solvent to remove any remaining grit.

From now on, be extremely careful to keep dirt, metal filings, and greasy fingers off the interior motor components. Set the components on a clean white cloth (diapers are great) and use a hair dryer or heat gun to completely evaporate all solvent. This is important because some solvents, notably alcohol, draw water from the air and this may later cause corrosion. After drying, check for residual grime with a jeweler's loupe or other magnifier. If there is any, remove it with a dry Q-tip.

Use a pin or toothpick to place a minute drop of light silicone-based gun oil such as Outers on the brush pivot points and the bushings. Work the oil into the bushings and remove any excess with a Q-tip. Before reassembling the motor, check the fine wires from the armature to the commutator segments. Broken wires here are almost impossible to correct, so if you find any, replace the motor.

To reassemble the motor, place the brushes and end plate back on the commutator and shaft. Slip a straightened paper clip through the motor housing bushing to hold the armature/brush/end plate assembly and slide the housing down. The motor magnets will attract the armature, so work carefully. Note that there is a brass ring inside the motor housing that accepts the screws which hold the end plate in place. The ring is free to rotate, so check to see that the holes in the brass plate and end plate align with the end plate and motor housing marks you made earlier. Mount the end plate with its screws, but leave it loose enough to be rotated.

Apply power and let the motor run for 5 minutes in each direction. It should run at the same speed in either direction. Motor speed is affected by the relative position of the magnets and the position in rotation at which the brushes touch the commutator segments. In fact, there are two positions, called null points, at which the motor stalls and will not run at all. Figure 6-6 shows the effect. A procedure suggested by Tom Gist ensures that the motor will run at the same rpm in both directions. Turn the motor end plate until the motor stops, then go beyond this point until the motor starts again, and finally back up until the motor stops once more. The null point is halfway between the two end plate positions where the motor first stops, then restarts. Turn the end plate 90 degrees from the null point. The original alignment mark should coincide closely. Now tighten the screws that hold the end plate.

Be careful when replacing a brass pinion. Position the back end of the motor shaft so that the shaft, not the internal seats of the armature, absorbs the pressure when the pinion is pushed on the shaft. Use a short length of brass tubing whose diameter is slightly larger than the shaft to press the pinion in place. If you have a drill press, chuck the brass tube in it and use the

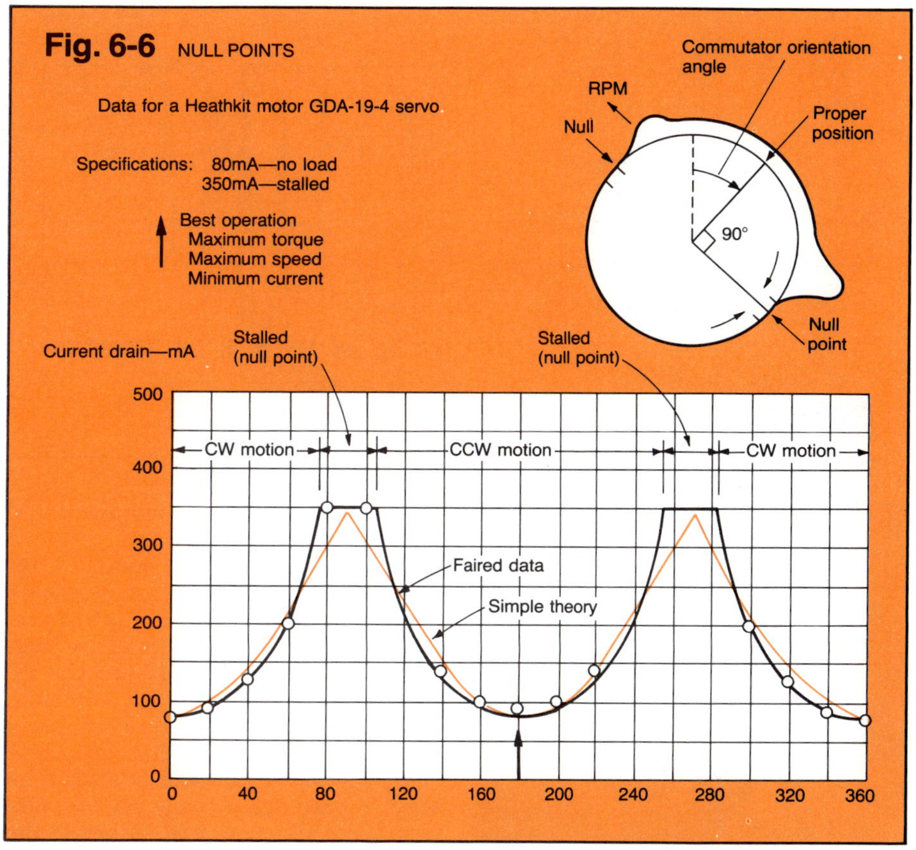

Fig. 6-6 NULL POINTS

This armature is filthy, but the brushes seem to be okay.

The armature and end plate are now free of the motor case. Precisely mark the location of the end plate relative to the hole in the brass retaining ring inside the motor case or later it will be very difficult to realign the parts.

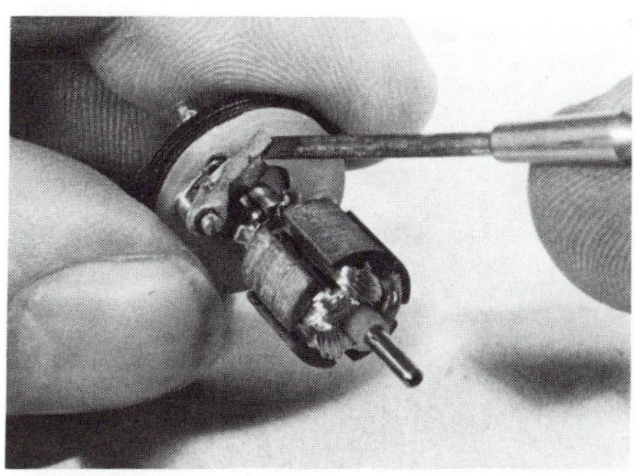

Another pair of hands helps here. Brushes must be carefully lifted clear as the armature is removed from the end plate. The little spring on the brush carries current to the armature.

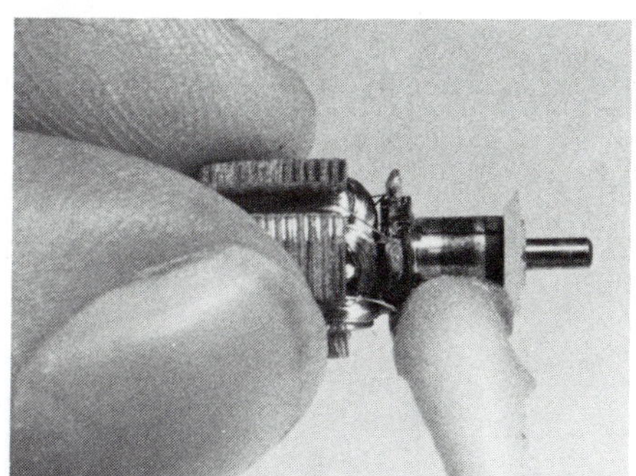

Q-tip and solvent remove most of the dirt from the armature, and burnishing with emery cloth will get the rest.

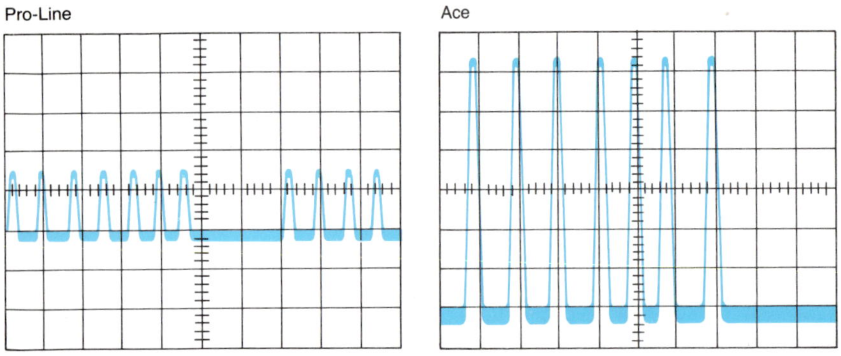

Fig. 6-7 TYPICAL RECEIVER WAVEFORMS

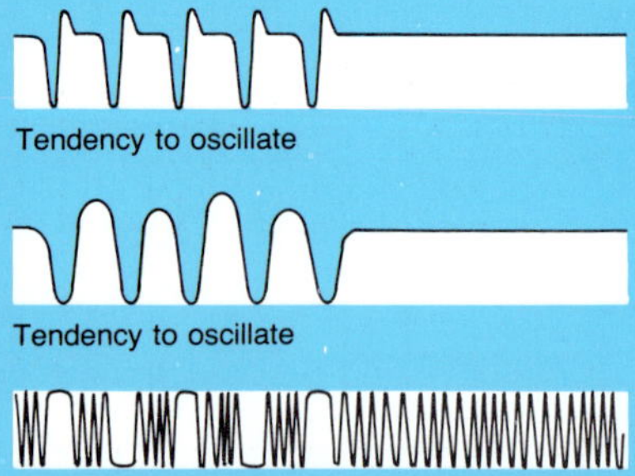

Fig. 6-8 UNSTABLE RECEIVER

handle that lowers the drill press chuck to control the pressure on the pinion as you press it in place.

All this is a lot of work, but if a new motor costs $10, and you have 10 servos to repair, a few winter evenings' work can save $100.

Step 8. Reassemble the servo by reversing the disassembly steps. Don't forget to replace any insulators or protective foam padding between the feedback pot and the PC board and the PC board and case. Plug in the servo and check for proper center. If offset, use a jeweler's screwdriver or hex wrench to reset the servo to dead center. The servo is once again ready for use. Obtain new grommets and reinstall the servo in the model.

Receiver

The receiver is the next most troublesome unit, accounting for about 10 percent of system failures. Most receiver failures result from broken power, servo, and antenna wires, or from broken wiring between decks of a two-deck receiver. Such failures are easily detected visually or by continuity tests with an ohmmeter. Replace broken wires, being certain to use 19-strand RC hookup wire. This is sold by Ace and other manufacturers. Check switches as well as wiring, replacing broken switches and cleaning dirty ones.

Much less frequently, the receiver fails because it has become detuned, most often as a result of a severe crash. Unless you have a schematic and an oscilloscope, just replace any broken parts with exact replacements, then send the system to a service center for re-tuning. Some kit-built receivers can be tuned without an oscilloscope using the technique explained in the last chapter. If you do have a schematic and an oscilloscope, begin re-tuning by removing the wax, silicone rubber, or other compound used to secure the ferrite slugs in the RF coils and IF transformers. Use a hobby knife or jeweler's screwdriver and work slowly to avoid cracking the slugs. Inspect the slugs for cracks and breaks. If you find any, discard the slug and order an exact replacement.

Consult the schematic and locate the ground (0 V) power connection. Locate the first audio amplifier transistor; it is usually associated with the output from the last (black) IF transformer. If the receiver has crystal filters, the audio signal may not appear until after one or more stages of amplification. Now connect a 10- to 22-μH choke from receiver ground to the oscilloscope ground clip. Connect a second 10- to 22-μH choke from the detector or tuning point just located. Turn on the transmitter and receiver, collapse the transmitter antenna, and place the transmitter on a workbench 6′ or 7′ away. Observe the detected output waveform on the scope; Fig. 6-7 shows typical waveforms. When tuning, it is important not to saturate the AGC with too strong a signal because the AGC will reduce the signal level of the preceding receiver stages, making it difficult to achieve good peaking. On the other hand, if the signal level is so low that receiver noise dominates, poor tuning also results. With the transmitter positioned as above, the detected audio signal should be a duplicate of the encoded pulse train and may be positive-going or negative-going. With a strong signal, the output baseline should be flat or nearly so and the output audio pulses should be clearly defined. Anything else indicates a problem in the AGC circuitry or some other difficulty more serious than detuning.

With the transmitter antenna still fully collapsed or, if possible, removed, move the transmitter farther away from the receiver until the baseline begins to display small amplitude noise spikes. Record the oscilloscope gain

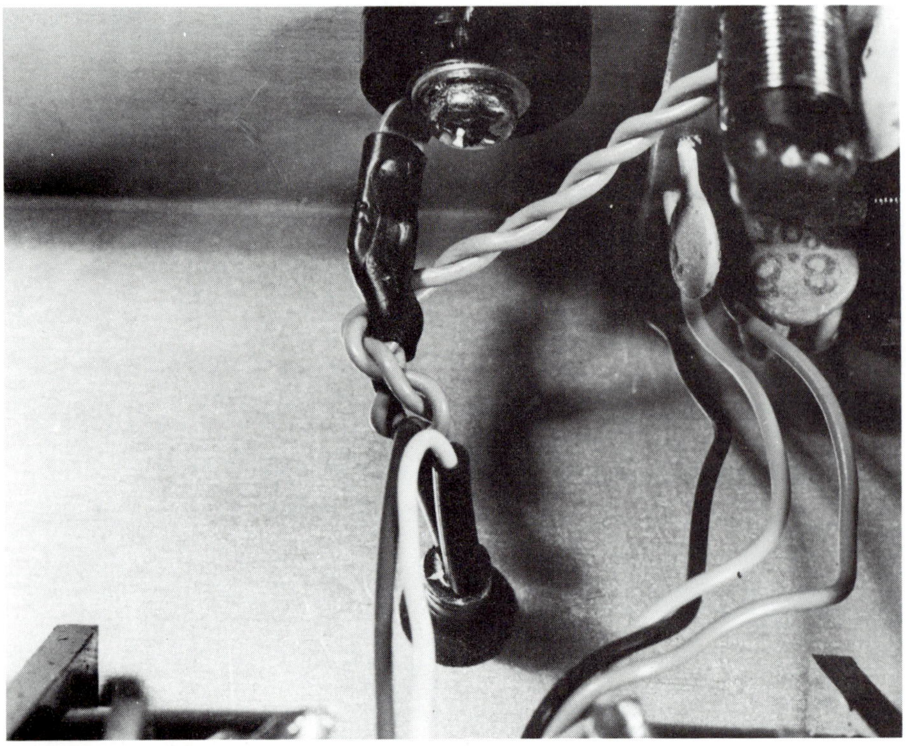

A look inside this transmitter revealed that the wiring from the RF deck to the antenna was twisted almost to the breaking point.

Fig. 6-9 ENCODER WAVEFORMS

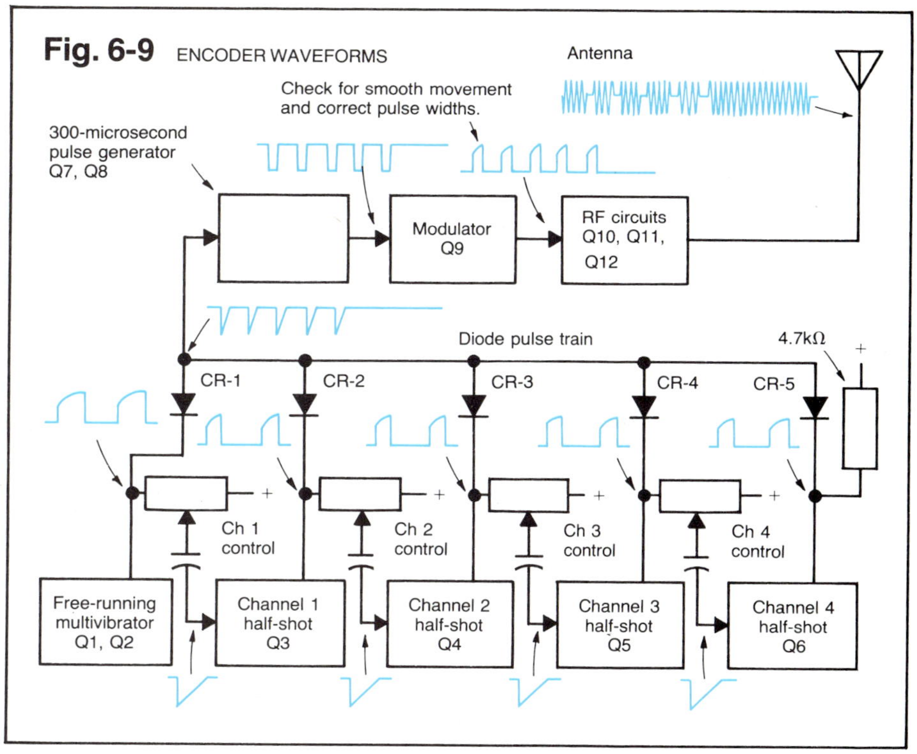

and signal amplitude for comparison after tuning. During the following steps, move the transmitter progressively farther away from the receiver as its sensitivity increases so that the same detected signal level will be maintained. Use a nonmetallic tuning wand to peak the IF transformers, going from yellow, to white, to black transformers in that order. Peak the fourth IF transformer if your receiver has one. Repeat as necessary. Do not turn the slugs very far as you may establish a false tune point. A half turn to a full turn is usually enough. Now tune the RF coil or coils (some receivers have only one, most have two, a few have three) in the same way, starting with the antenna tuning coil and progressing to the mixer tuning input. Repeat as necessary, because there may be some interaction between the coils.

At no stage of tuning should the baseline be allowed to return to its fully flattened condition, since that means the receiver is nowhere near its maximum range (sensitivity). The receiver's range with the transmitter antenna off will vary widely from system to system and from band to band. Antenna-off range increases at the higher frequencies because the receiver antenna's length comes closer to being a quarter wavelength. Range generally doubles from 27 MHz to 53 MHz and increases again from 53 MHz to 72 MHz.

After the tuning adjustments have been made, check for stability. Short the crystal pins with your finger. If the local oscillator stops, it should restart immediately when you remove your finger. Turn the transmitter off; the baseline on the scope should sit quietly, showing only random noise. Any cyclic activity is a sign of instability in the receiver that is best corrected by a serviceman, Fig. 6-8.

Now secure the RF and IF tuning slugs with a few drops of molten candle wax. Put the receiver back into its case and range-check with the transmitter antenna off, collapsed, and extended. Record these ranges and the date in your system's log so you'll be able to make comparisons later.

To summarize, most receiver problems are caused by loose or broken wires — inspect the receiver frequently and replace broken wires. Do not attempt to re-tune a manufactured receiver unless you have a schematic and an oscilloscope.

Transmitter

Transmitters seldom need maintenance other than routine inspection for broken wires, damaged antennas, and antenna mounting hardware. Let's assume, however, that you've been given a system that's 5 years old and that needs a complete overhaul. You've already replaced the batteries in the transmitter and airborne units, and you've serviced the receiver.

Before you begin the overhaul, be aware that you cannot re-tune the RF section or change the crystal unless you have an FCC 2nd Class license. There are only two exceptions to that statement: You may change crystals on 27-MHz units if they have plug-in crystals and are designed for owner frequency change; and if you're an amateur radio operator with at least a General license, you may re-tune transmitters that operate on the 53-MHz ham band. Do not attempt to replace the RF amplifier if its internal capacitance makes up part of the RF amp output tuning. If it does, let a service center make the replacement.

The most probable failures in an old transmitter will be the antenna connection wire and the wires to the control pots. Look for broken wires, twisted mounting hardware that has shorted insulation, and loose parts. Antennas are relatively inexpensive; if a new antenna is needed, use an exact replacement. Replace wires to control pots if the insulation has cracked or has become too stiff for easy movement.

In an extreme case, you may decide

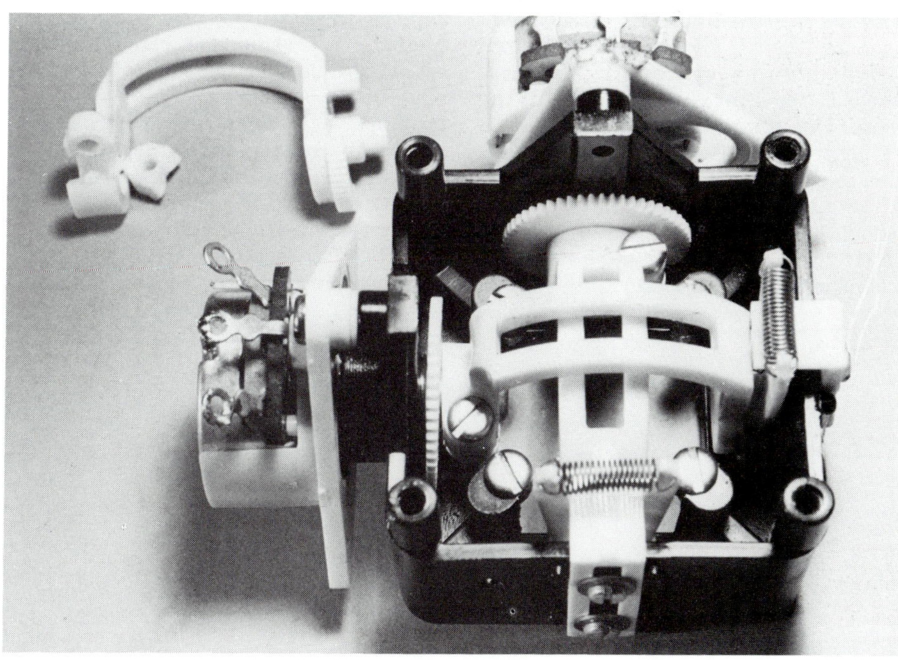

A broken stick bail is easily replaced.

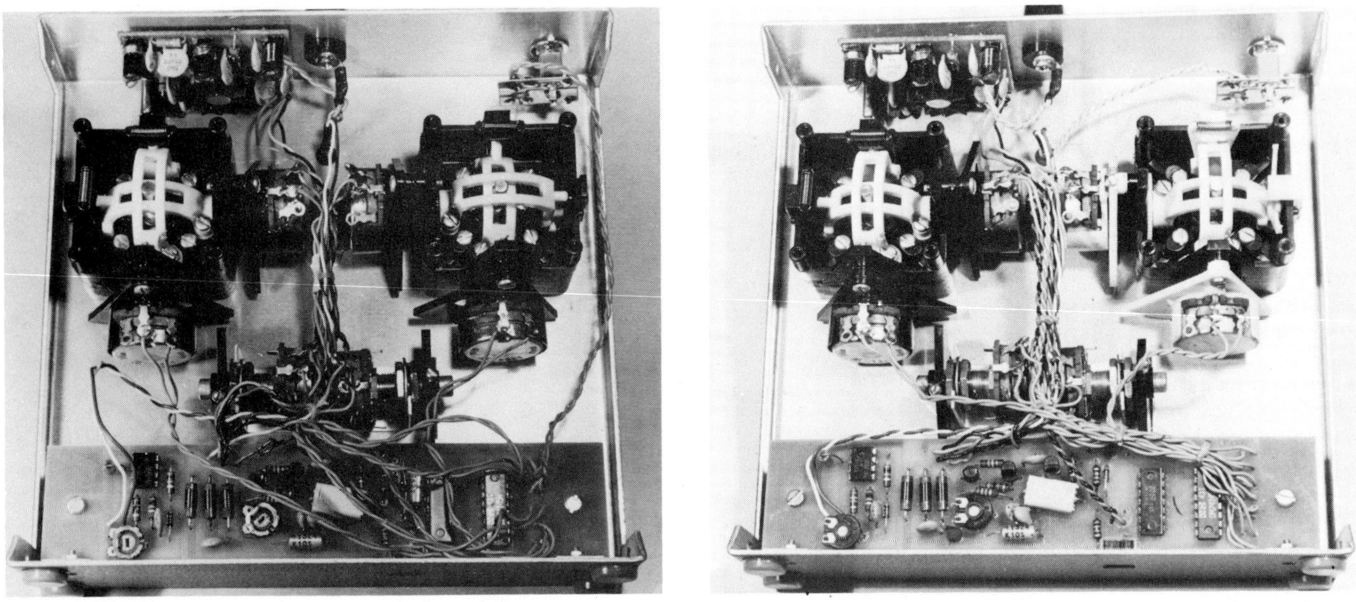

(Left) Maintenance sometimes borders on rebuilding. This kit-built transmitter was sloppily wired, so I decided to replace all wiring harnesses. (Right) New wiring looked more workmanlike and offered greater security against breaks. A new stick bail, new sealed potentiometers on the encoder board, and a new antenna brought the system back to health.

to replace all control pot wiring and all wiring harnesses. Before doing so, check the control pots for noise and nonlinearity. Hook up the oscilloscope ground to 0 V and put the test probe on the encoder so you can see the series of control pulses. Move each control and trim to see that their respective pulses move smoothly and linearly on the scope, Fig. 6-9. Some scopes have a magnification of horizontal gain feature that allows a given channel to be displayed over most of the scope face. This permits a real "vernier" look at the individual channels. If any channels are rough or badly nonlinear, replace their control pots.

Replacement of sealed pots, as some systems use, forces you to decide whether you will order a replacement pot from the RC system manufacturer, or whether you will buy replacement parts from a local supplier. Buying from the system manufacturer is easier, but more expensive. If you decide to replace only the worn parts, disassemble the stick assembly, carefully noting the flat index mark on the pot shaft that indicates where it mates with the stick bail. Determine the pot make and part number as well as the shaft size, body diameter, and the length and diameter of the threaded pot mounting shoulder. Armed with these facts, buy replacements for worn parts.

Rewiring involves disassembly of the transmitter. Clean each piece as you go with a solution of dishwashing detergent and warm water, scrubbing away stubborn dirt with an old toothbrush. As you remove wiring, make sketches and tag wires with colored tape or ties so that you'll be able to get everything back to its proper place. Make sketches also as you disassemble stick assemblies. Check sticks for cracked or broken parts. Remove PC boards, check component values, and replace damaged or degraded components. If you replace the RF amplifier transistor, be sure to remount its heatsink (if any) after replacing the transistor. Check the PC board itself for cracks or broken lands. Inspect solder joints for crystallization and cracks.

As you begin to rewire, replace the power switch. If you've gone to all this trouble, there's no use leaving the old one in there. Rewire the power wiring first, but be sure to cap the ends of the live power wires that will be connected to the RF deck and the encoder. Follow the instructions in the last chapter, and fabricate a new wiring harness. Use sleeving and waxed cord or nylon ties wherever it was used originally. Cut all cables to the same length as the old ones. Wire stick assemblies before they are reinstalled. When all wires to switches, sticks, and connectors are in place, route the wires in the power supply cables to their proper junctions on the PC boards. Solder these wires. Scrub the solder side of the encoder PC board with a toothbrush wetted with denatured alcohol, let dry, and spray coat both sides of the board with Krylon or GC PC board coating. When the coating has dried, remount the board and check over the entire job. The transmitter should be as good as new for several more years.

This concludes my discussion of maintenance procedures for RC systems. You'll have noticed that while some steps require an oscilloscope, most do not, and that visual inspection is the most frequently used maintenance technique.

7 Troubleshooting

You should know enough about troubleshooting to be able to identify which part of an RC system has caused it to fail. Having identified the defective part, you may choose to have the fault repaired by a professional service technician or to make the repairs yourself.

Interference

Whenever a system behaves erratically, check for interference before looking inside the system. If you're at an operating site, inquire (politely please) whether anyone has turned on to your frequency. If no one has, the problem may stem from illegal CB operation on the 27-MHz band, or from various industrial users on the 72-MHz band. Switch off the receiver and wait a few minutes for the interference to stop before trying again.

If the model has suffered a minor crash which you think may have been caused by interference, first inspect the servos to see if they are at or near center, then examine the receiver, servo, and battery wires for breaks. Check that the receiver antenna is intact. Unplug all servos except one that is properly centered. Disconnect the pushrod or other linkage to the servo. Turn on the transmitter. Turn on the receiver but be ready to turn it off quickly if the servo jumps about. If the servo snaps to attention at neutral and follows the stick properly, the crash probably was caused by interference. Plug in the remaining servos; if they all work properly, interference almost certainly caused the crash.

Before returning to operation, remove or collapse the transmitter antenna and range test the system. Random or cyclical servo motion as the system approaches maximum range may indicate continuing interference. Walk back to the model, leave the receiver on, and turn off the transmitter for a few minutes. That's the only time you're allowed to turn off the transmitter without first turning off the receiver. If the servos sit quietly, there was no interference or it has stopped. If the servos jump around, the problem was most likely caused by interference which has continued. The erratic servo movement may also be caused by a battery pack that has gone bad. Get out an ESV or VOM and check the transmitter and receiver battery pack voltages under load. If the receiver battery pack has gone bad, and you have a fully charged spare pack in your field box, substitute it for the bad pack, check servo motion, range test, and resume operation. If the transmitter battery has gone bad, substitute a good pack. If the problem does not disappear when good battery packs are substituted, the system has more serious problems with the transmitter, receiver, or servos, so you might as well go home and worry about it later.

Basic troubleshooting

To illustrate the sort of troubleshooting you can do with little or no test equipment, let's assume a worst-case situation: The set has suffered a catastrophic in-flight failure and your precision-scale Hyper-Apple-Knocker has planted itself squarely in the middle of the flying field. Collect yourself and calm down. Remember I said that no man-made system is completely reliable. In order to prevent the servos from driving off-center due to noise or interference, leave the transmitter on and do not collapse the antenna. Do not move the transmitter sticks as you approach the wreck — if servos are damaged or linkages are jammed, additional damage may result from moving the servos. Reach in and turn off the receiver. Carefully disconnect the receiver battery pack, and only now turn off the transmitter. Search the area for connectors, output wheels, servo tray parts, and other small pieces that may have broken off. Pick up the model and all the pieces and take them back to your field box.

Do not attempt to operate the radio at the field; the system's going to have to be thoroughly cleaned and checked before operations resume. At home, remove the equipment from the model and separate the airborne pack into its constituent servos, receiver, battery pack, and wiring harness. Clean all dirt and oil from the components. Inspect the wiring harness for breaks, bare wires, shorts, and other damage.

We now want to find out what quit, why, and what damage, if any, has been done to the set. Figure 7-1 presents a simple troubleshooting tree. Such trees are used by maintenance personnel to trace and correct problems in full-size aircraft. First check to see if the transmitter is radiating. An RC transmitter produces RF carrier harmonics or modulation sidebands that can be detected with an FM radio or TV. Extend the transmitter antenna, hold the transmitter near the TV or FM radio, and tune the TV or FM radio until you hear the buzz of the transmitter on the radio or see lines on the TV. Adjust the TV's vertical hold until the lines stop moving across the screen.

A well-equipped repair station at Royal Electronics. Signal generator at left is essential for a service technician but too expensive for most amateurs.

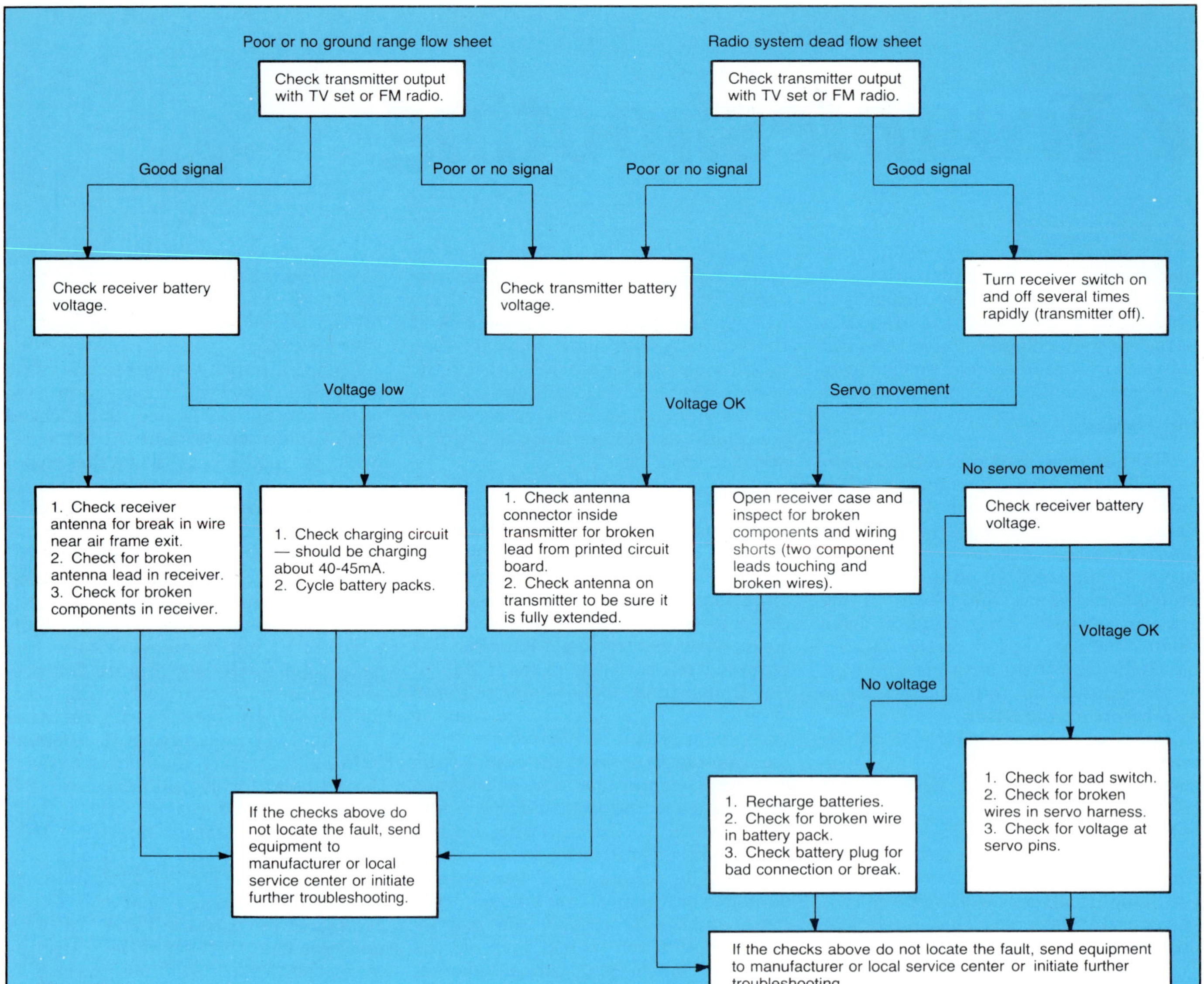

Fig. 7-1 TROUBLESHOOTING TREE

Then move each transmitter control and observe that the lines change position. If you can't find the signal on any of the VHF TV channels or anywhere on the FM band, the transmitter isn't radiating. Possible causes for transmitter failure include broken switches and wires, bad batteries, and damaged components on the PC board. Fix these using the techniques in Chapter 6.

If the transmitter is radiating, check the receiver. The crystal, the receiver's most fragile component, may have broken in the crash. Shake the receiver, listening for a rattle from the crystal. If it rattles, replace the crystal. If you don't hear a rattle, but suspect the crystal may nevertheless be broken, wrap several turns of the receiver antenna directly around the transmitter antenna. This may apply enough RF to "drive through" the crystal into the rest of the receiver. If the receiver now works, replace the crystal. If the crystal seems okay, but the receiver still doesn't work, check its IF stages. Turn on the receiver and hold its antenna close to a monitoring receiver. The local oscillator in the receiver radiates RF at a wavelength 455 kHz removed from the transmitter frequency. Thus, your receiver should quiet the monitor at a point 455 kHz above or below the transmitter frequency. If it does not, the receiver needs to be repaired. If the battery packs, connecting cables, transmitter, and receiver check out, but one or more servos operate erratically, obviously the servos need work.

These are the only troubleshooting techniques you can undertake without fairly sophisticated test equipment. They do allow you to repair most physical damage and allow you to tell a service technician which part of the system has failed.

Test equipment

More advanced troubleshooting requires:

- A digital voltmeter (DVM), FET multimeter, or VOM with an input impedance of at least 20,000 ohms per volt. Inexpensive (less than $35) multimeters are adequate for battery checks, but should not be used for circuit measurements because their input impedance is too low. If you don't have a good multimeter, you can use a VTVM, though this can't measure amperage.
- A triggered oscilloscope with the capability to display signals as low as 50 millivolts amplitude and at bandwidths up to 10 MHz.
- An RF/audio monitor, RF signal strength meter, or one of the demodulator probes in Fig. 7-2.

Optional equipment includes:

- An RF signal generator. Unfortunately, inexpensive RF signal generators are too leaky and too poorly calibrated for accurate tests of receiver performance. A signal generator that will do the job costs at least $1500, unless you can find and modify surplus military equipment. Given the high prices, most hobbyists do without a signal generator.
- A dip meter. This is useful for

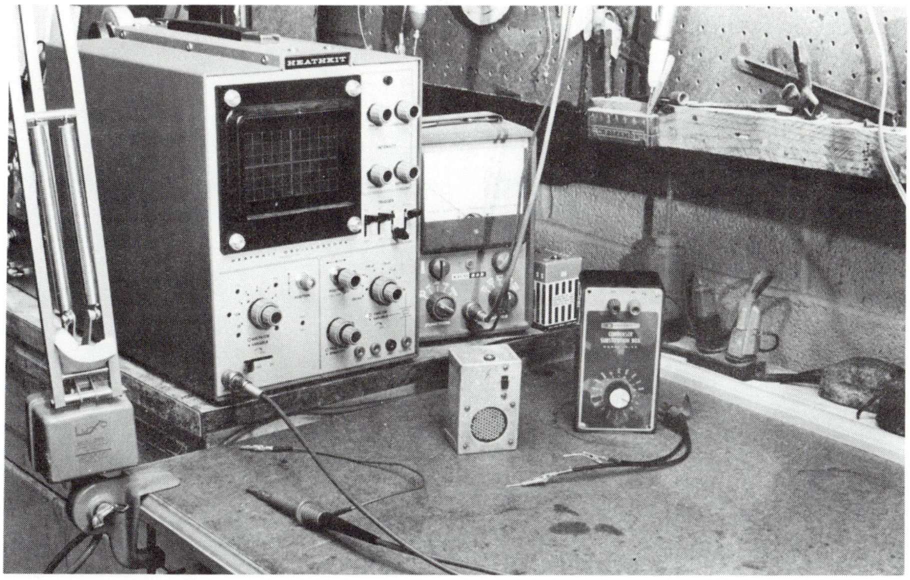

Test equipment lasts for years. This kit-built gear dates from 1972.

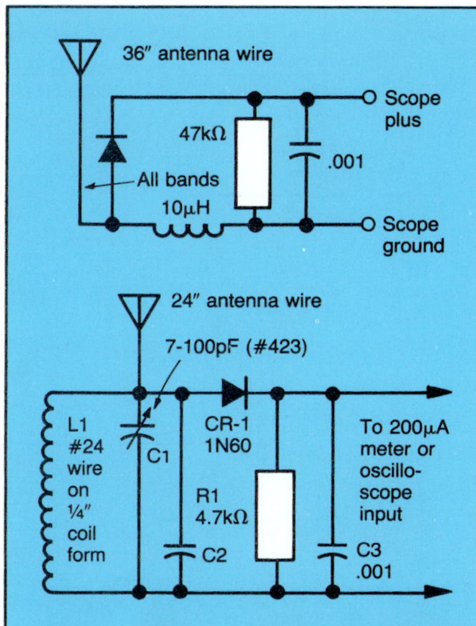

ITEM	27MHz	53MHz	72MHz
L1	7½ turns (open wound)	7½ turns (open wound)	2½ turns (open wound)
C2	56pF	OMIT	OMIT

Fig. 7-2 DEMODULATOR PROBES

checking coils. Heathkit manufactures one, Model HD-1250, with a range of coils for good frequency coverage.

- A capacitance test meter. This allows you to check capacitors before and after they are installed in a circuit.
- A transistor checker. Transistors are among the less reliable electronic components in RC systems, so a transistor checker is useful.
- A crystal checker. By no means a necessity, consider a crystal checker only if you plan extensive service work.
- Battery cyclers and servo testers are not test equipment as such, but are convenient servicing aids.

If you're an amateur radio operator, an electronics hobbyist, or a professional electronics technician, you probably already have this test equipment. If you do not own this equipment, think seriously before beginning to acquire it. The money you'll invest will far exceed the money you would save by not having to send broken sets to a professional serviceman. It's a justifiable expenditure only if you enjoy working with electronics and are willing to write off the expense just as you do other costs of pursuing your hobby.

Information

In addition to test equipment, you'll need schematics and normal values for the sets you troubleshoot. Kit manufacturers provide this information in their assembly manuals, but it may be difficult to obtain for other systems. Unfortunately, there is no equivalent to the Sams Photofacts for RC systems. Photofacts give a service technician schematics, waveforms, normal voltages, and alignment instructions for thousands of radios and televisions. If you have the time and a schematic, but no other information, you can develop the equivalent of a Photofacts for your system. Figure 7-3 shows the sort of information you should record. Use a DVM, VTVM, or VOM with a high input impedance to measure the DC voltages throughout the transmitter, receiver, and servos both with and without a transmitted signal present. Use an oscilloscope to measure AC waveforms, sketch or photograph these, and record their amplitude, shape, and time base. Measure encoder waveforms with the controls at both neutral and full-throw positions. Unless you have a sophisticated oscilloscope, you won't be able to observe the transmitted RF waveform. However, you can use the probes in Fig. 7-2 to demodulate the audio frequency waveforms in the receiver. Set the scope for a given demodulated pulse amplitude and record all the data. Also write down the exact positions of the transmitter and receiver so you can put them in exactly the same places whenever you repeat this procedure. Any significant deviation in the future from the recorded values indicates a problem. Record receiver waveforms with the transmitter off, with the transmitter on and close enough to the receiver to saturate the AGC, and with the transmitter antenna removed or collapsed and at the maximum range at which the servos still respond. The last condition corresponds to maximum receiver sensitivity, and, for an AM receiver, AGC at full gain. In an FM receiver, it corresponds to loss of capture and a drop below the limiter level for the demodulator. At maximum range, several of the traces may have amplitude too low for your scope to measure. Accurately record all receiver waveforms because subtle changes are as indicative of problems as a new "plip" in an electrocardiogram.

Although AGC dynamic range and linearity can only be checked with an accurately calibrated RF signal generator, you can obtain a rough estimate of AGC performance with a VOM. Because the strength of the RF field radiated by the transmitter decreases, theoretically, with the square of the distance from the transmitter, you can measure AGC characteristics by positioning the transmitter 2 feet, 4 feet, 16 feet, and so forth, from the receiver. The AGC line should display a relatively linear change in DC level throughout its dynamic range as RF signal strength linearly decreases when the transmitter is moved farther from the receiver. Record all values, and be alert for changes in the future.

Advanced troubleshooting

Assuming that you have an array of test equipment and schematics, and know the system's normal characteristics, you can further troubleshoot the system. Start with the encoder. Check the modulation waveform that is routed to the transmitter RF section. Is it stable? Are all pulses (one more than the number of channels) there? Is the amplitude correct? Is the sync pause of the proper duration (usually around 5 ms)? Are the control pulses of the specified duration (about 1.5 ms)? If all these are correct, the encoder is okay. If the encoder is completely dead, check for proper voltages throughout. The voltage regulator, if any, may be dead. If supply voltages are correct, check the clock circuit. If the clock is running but one or more channels are missing, suspect failure of a one-shot or half-shot. Keep in mind that the channel

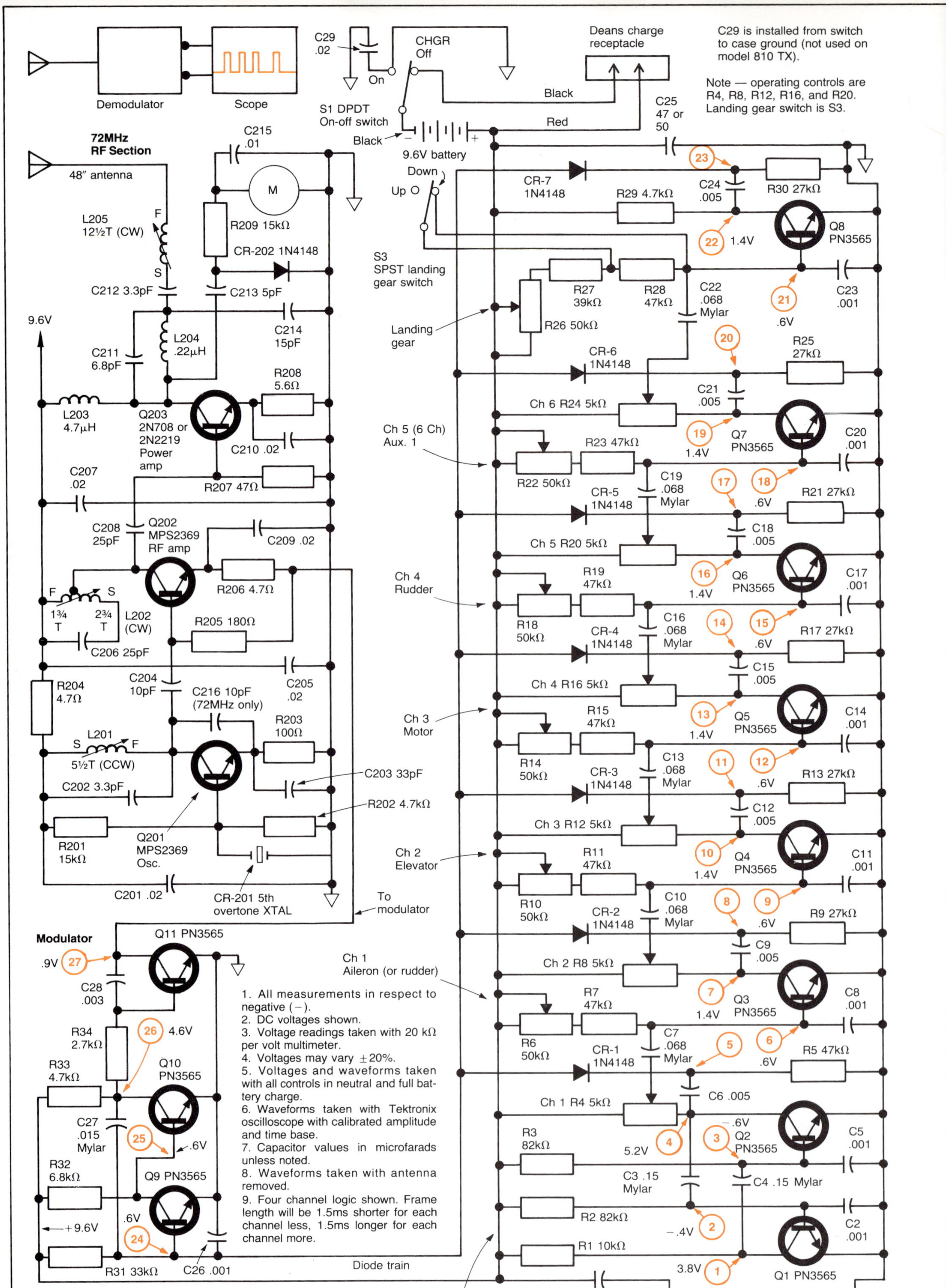

Fig. 7-3 RECORDING NORMAL VALUES

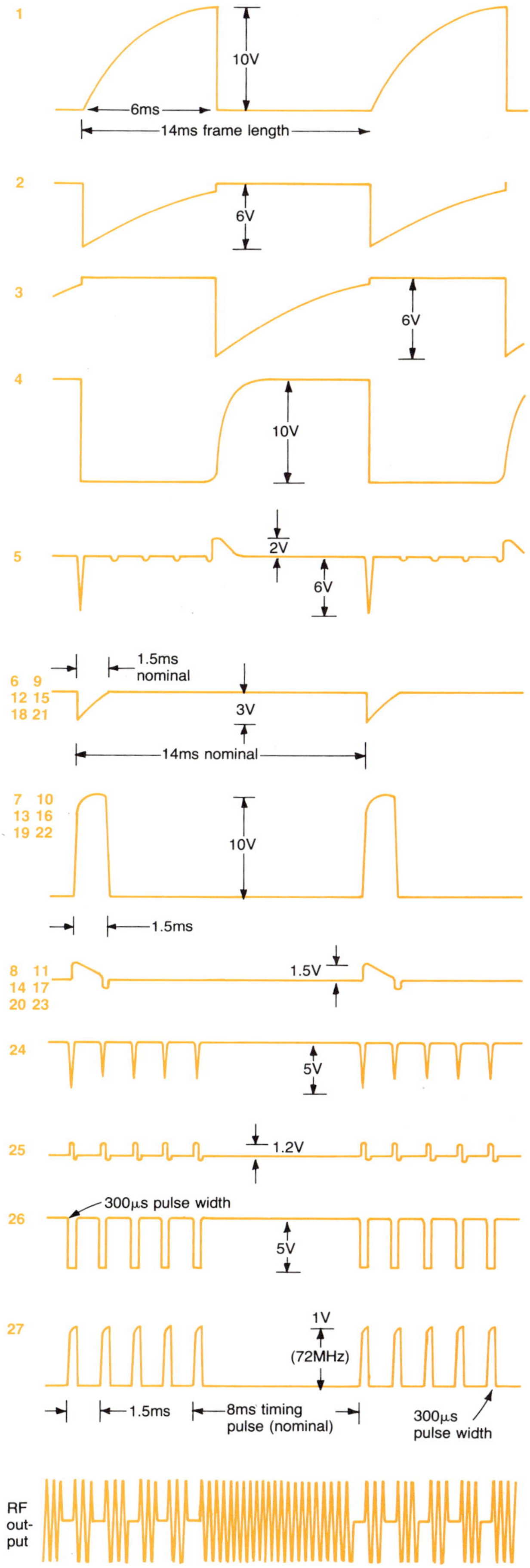

Four channel logic shown. Subtract one pulse width for each channel omitted; add one pulse width for each extra channel above four.

A simple transistor checker is a big help when testing suspected components for shorts, opens, and other defects.

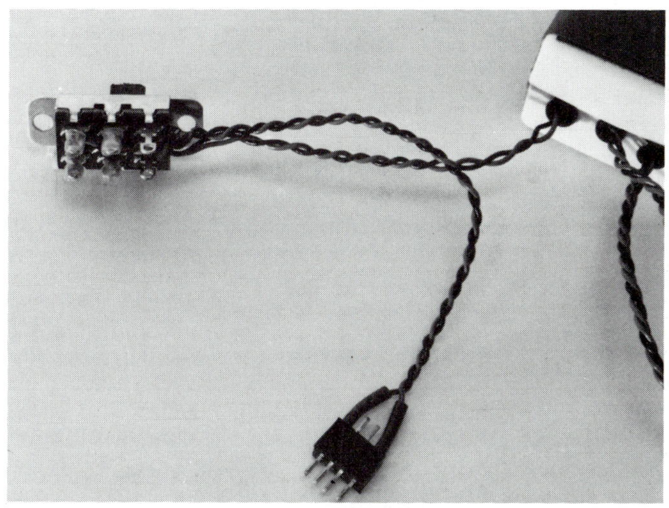

Check the wiring on switches for breaks and crystallized solder joints. Replace excessively dirty switches.

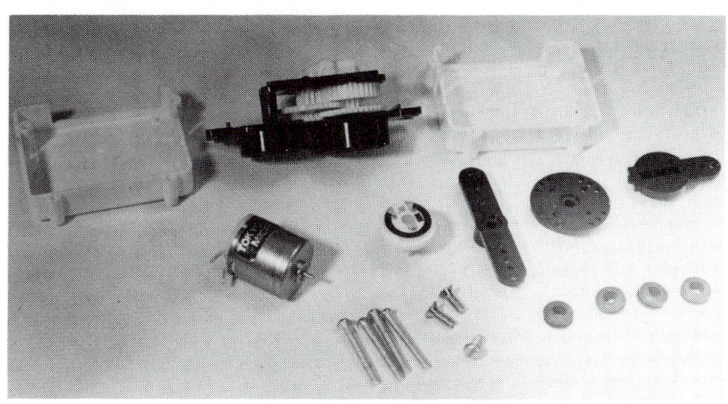

You can replace any part of a worn or crash-damaged servo.

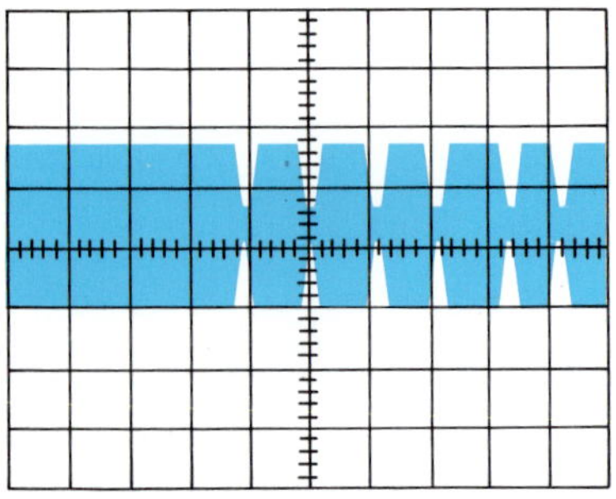

Fig. 7-4 AM MIXER OUTPUT WAVEFORM

pulse generators are usually in series and are sequential. If your set has a commutated encoder, look for an open or shorted input circuit.

Like the dog that caught the sports car, "What are you going to do with it now that you've got it?" At this point, you'll have to substitute components known to be good for those suspected to be defective. Those most likely to be defective are, in declining order of probability: transistors, diodes, ICs, capacitors, and resistors. If the failure is intermittent, poke at suspected components with a tuning wand or some other nonmetallic stick to see if the problem comes and goes. If you suspect the intermittent fault is caused by vibration, mount a blunt piece of heavy solid wire in the collet of an electric engraving tool and use it to transmit vibrations to the suspected components. Be certain that the engraving tool isn't creating electrical interference that makes the test invalid. If the difficulty seems to be temperature dependent, freeze suspected components with an aerosol cooling gas, observing changes as the part freezes, then warms up.

Next troubleshoot the receiver and decoder. First check all DC voltages to reveal defects in the voltage regulator or elsewhere. If necessary, fix the voltage regulator before proceeding further. If the bandwidth and sensitivity of your oscilloscope are sufficient, check the local oscillator. If not, move on to the mixer output. With the transmitter antenna fully extended and near the receiver, you must see a waveform similar to Fig. 7-4 for an AM set. An FM mixer will not display amplitude modulation, but you should see a shift in frequency where an AM pulse would be seen in an AM receiver. The mixer output is the intermediate frequency of 455 kHz, so should be easy to see on any good oscilloscope. Should there not be proper output from the mixer transistor (in a discrete component receiver), replace the mixer transistor. If this doesn't solve the problem, replace the local oscillator transistor, then the crystal, and beyond that, the associated capacitors and resistors.

Now trace the signal through the IF strip. If the signal disappears at a transistor, IF transformer, or some other component, replace that component. If you have a signal generator, an injected IF signal at the mixer output makes tracing the IF signal easier. The AGC in an AM receiver may be faulty. This can clamp off all IF stages, and if the AGC acts on an RF amplifier, may affect that too. Check if the problem originates in the AGC circuitry by lifting a resistor in the AGC line or by cutting the AGC line on the PC board near the detector output. Place a trim potentiometer at that juncture with the ends connected to 0 V and the receiver supply voltage. Connect the pot's center tap to the AGC line. Vary the pot setting to see if the IF stages react. The point of failure shows up where the IF signal disappears. Replace transistors first, then IF transformers, then the associated capacitors and resistors. When you have regained an IF signal, reconnect the AGC to verify proper operation. If the IF disappears when the AGC is reconnected, replace the detector diode, AGC filter capacitors, and the AGC resistors in that order. The detected output for both AM and FM receivers is an easily traced set of audio-frequency pulses. Replace the component wherever the pulses stop and move on through the receiver to the decoder. Since most decoders are ICs, the signal will almost certainly disappear at one pin of a failed IC. Replace the IC. It's much less likely that a diode or capacitor will have failed, but check these if the new IC doesn't solve the problem.

With the transmitter and receiver checked out and repaired, it's time to inspect the servos. Any failed servo should first receive the maintenance procedure explained in the last chapter. Then check to see that servo amplifier waveforms are present. If they're not, the IC has probably failed. It's much less likely that a capacitor or resistor has gone bad. It's hardly worthwhile to spend a lot of time troubleshooting and repairing a servo amp. I usually buy a complete servo amp kit from Ace, Cannon, or Royal, and replace the whole thing. If you must repair an old amp, replace the IC, capacitors, and resistors, in that order, until the amp operates properly. Before reassembling the servo, check the servo motor's resistance and current levels as described in Chapter 6. After reassembly, check for servo linearity. Nonlinearity of up to 5 degrees is normal with all servos except those using the NE544 IC, which should give almost perfectly linear response. However, you should be concerned if the servo output arm moves more than 5 degrees more in one direction than the other because this may indicate problems in the servo or elsewhere in the system.

The battery charger is the easiest unit to troubleshoot. Most contemporary RC systems use a transformer-isolated charger that produces an output for the receiver or transmitter battery pack, or both. The transformer output is AC at about 12 to 18 volts peak-to-peak. This passes through a diode and a bridge rectifier. A large capacitor may be used to even out the waveform. In addition, a voltage-dropping resistor and an indicator lamp or LED are usually part of the circuit. Troubleshooting is simple: Use the oscilloscope to trace the waveform through the circuit, replacing defective components wherever the waveform stops.

After you've completed repairs or the repairs have been made by a technician, don't install the system in your rebuilt Hyper-Apple-Knocker for its first flight. Put the system in a stable old crate and make several flights to be sure that all defects have been repaired. Finally, cultivate a becoming modesty about your troubleshooting abilities. Don't work on someone else's system until you are truly proficient.

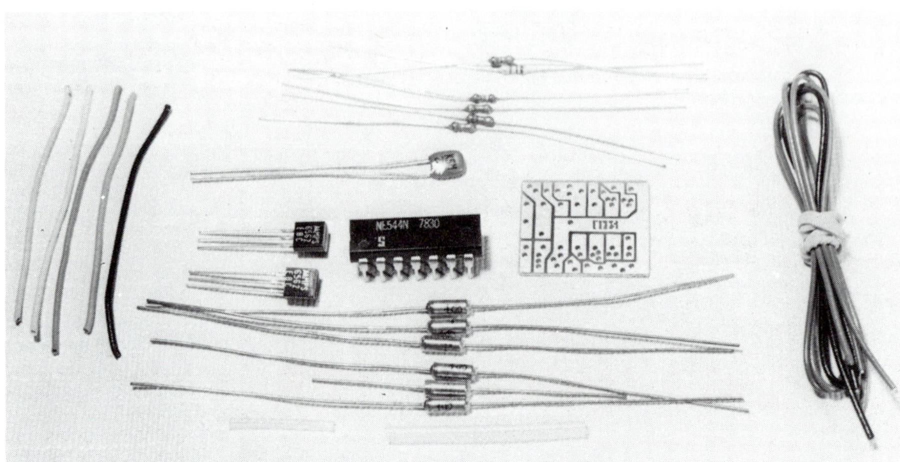

Servo amplifier kits are less expensive than repairs, so I usually discard a failed servo amp and install a replacement.

8 Installations

When I began RC modeling there were no special items for the installation of equipment in the model. We bought connectors at the local radio or TV store. The power switch was almost as large as a small servo now is and weighed more. Pushrods (or torque rods, as they usually were at that time) were laboriously bent from music wire. Pushrod keepers were short lengths of music wire bound to the pushrod with copper wire and solder. Now, just 20 years later, 14 pages are devoted to installation aids in the 1980 edition of the *Radio Control Buyers Guide.* It is not this chapter's purpose to cover the myriad installations possible in airplanes, cars, boats, helicopters, and the like. Several books already cover that subject. In addition, each modeler is an innovator who will find new ways of doing things. The purpose of this chapter is more to note what not to do in installations and to point out the common pitfalls to be avoided. I've seen experienced modelers make poor installations, and I've made a few that could have been better. Why? Usually because of impatience. The right control accessory wasn't available at the moment, so rather than wait for a trip to the hobby shop, the installation was jury-rigged. Another cause is plain laziness, not taking pains to do it right. Fine, if there's time to do it over, there was time to do it right the first time.

Rules

There is a set of rules to follow, without exception, in making RC installations. It is largely insignificant how these rules are satisfied as long as they are satisfied. The rules are:

- Plan the installation before you begin to assemble the model.
- Have all accessories on hand at the start.
- Make no installation that requires the servo to draw more than 150 to 175 mA with the model suspended in air.
- Provide adequate vibration damping for the receiver, battery pack, servos, and power switch. Protect the receiver, battery pack, and servos from impact damage.
- Protect all components from fuel, exhaust residue, dirt, and (in boats and seaplanes) water.
- Minimize servo wear caused by oscillatory loading of control surfaces. Oscillatory loading refers to unwanted forces such as flutter and cavitation.
- Protect all wiring and connectors from pinching, chafing, and breaking.
- Make sure all controls operate properly before, during, and after installation.
- Provide access to all components for maintenance. All components must be easily removable.
- Use servo mounting trays whenever possible.
- Tape or otherwise secure all connectors and wiring harnesses.
- Route the antenna wire as far as possible from servos and all wiring. Never alter the antenna length.
- Don't allow pushrods or other link-

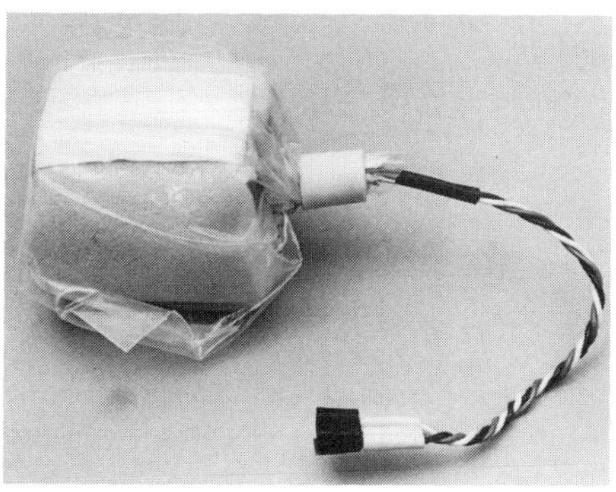

Foam rubber wrapped around airborne components protects them against engine vibration. A plastic bag over the foam-wrapped components keeps out dirt, fuel, and exhaust residue.

The RC gear in this car model had to be mounted with servo tape because there was no room for servo trays.

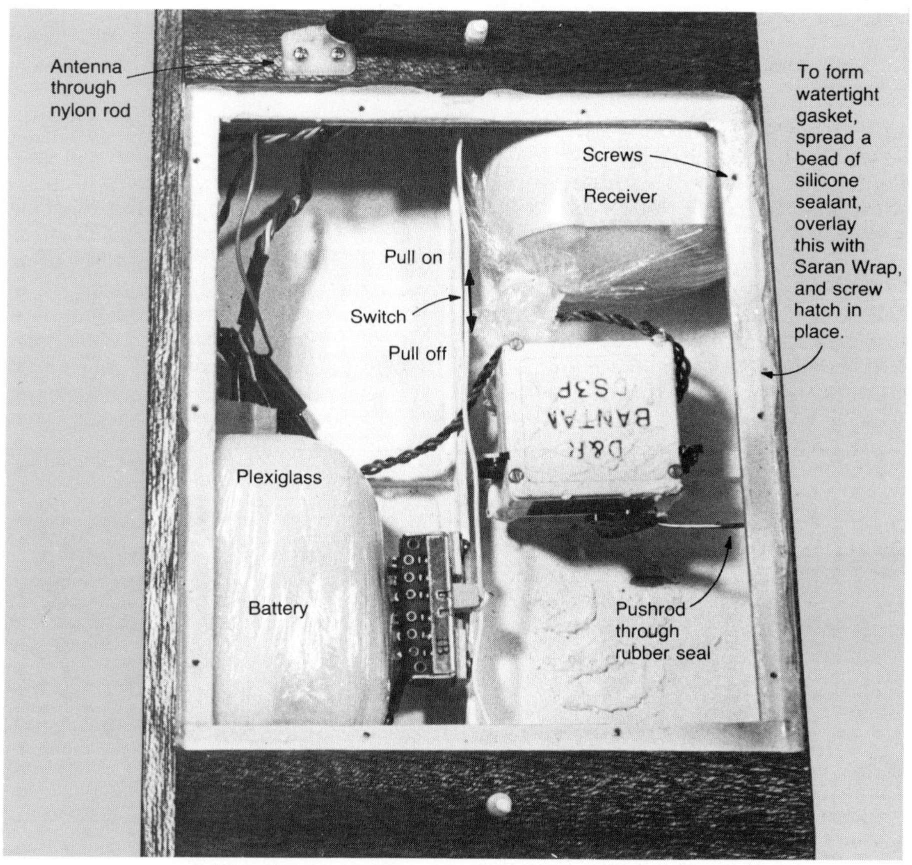

This boat once turned over and drifted ashore but the homemade watertight box protected the RC equipment from damage.

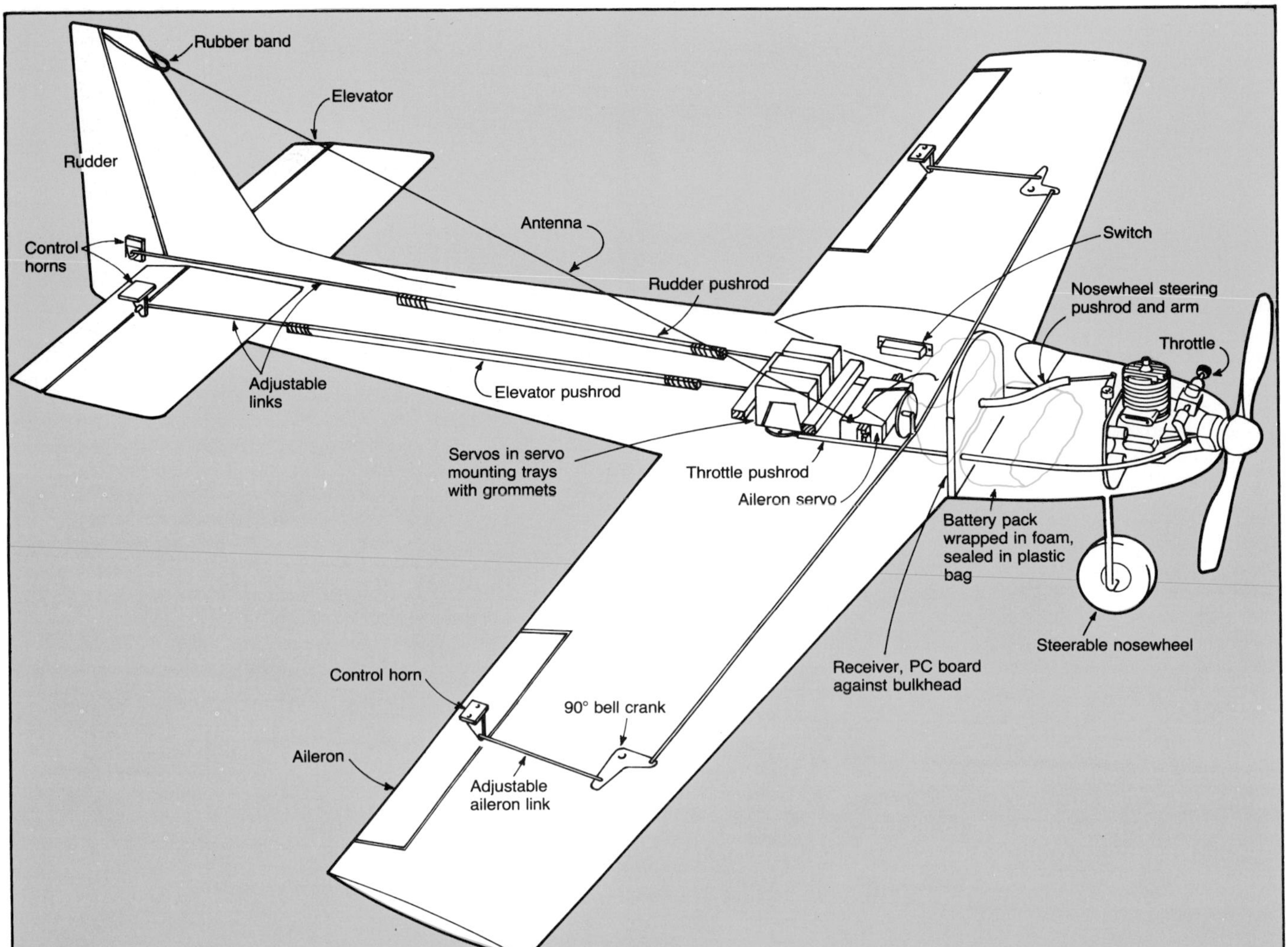

Fig. 8-1 CORRECT INSTALLATION

ages to rub against the receiver, battery pack, or one another.

Figure 8-1 shows an installation that conforms to these rules. Study it carefully.

Planning

If you plan the entire installation before starting to build the model, you will avoid the two most common installation mistakes: Failure to provide for passage of pushrods, and failure to take the direction of servo output motion into account. The locations of holes for pushrods or cable sleeving must be determined in the early stages of modelbuilding and the holes must be cut or drilled at the appropriate time. It is difficult to cut passage holes in a bulkhead in an assembled fuselage. Determine the direction of motion of each servo's output arm before locating servos. Plug each servo into the same channel output on the receiver. Exercise each servo, noting its output direction. Record that information in the system's log, and write it on the servo case, using the notation CW (clockwise) or CCW (counterclockwise) for the direction of rotation as viewed from the top of the servo output arm. You may want one or more servos to rotate in an opposite direction from the rest.

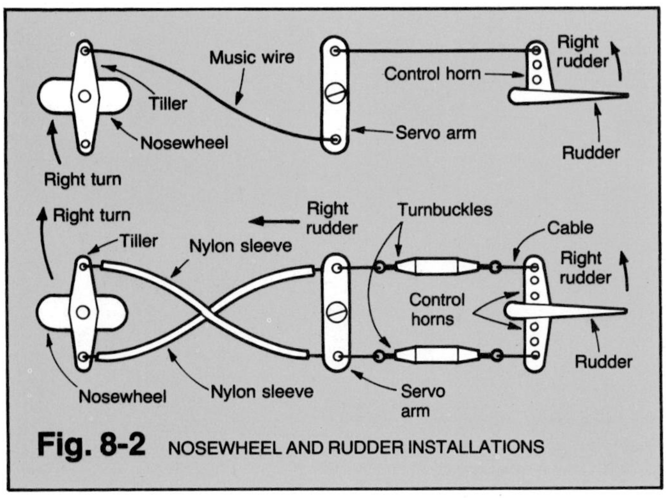

Fig. 8-2 NOSEWHEEL AND RUDDER INSTALLATIONS

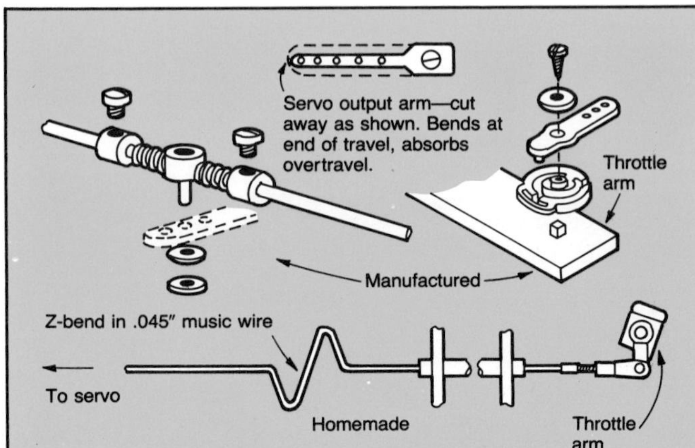

Fig. 8-3 THROTTLE OVERRIDE DEVICES

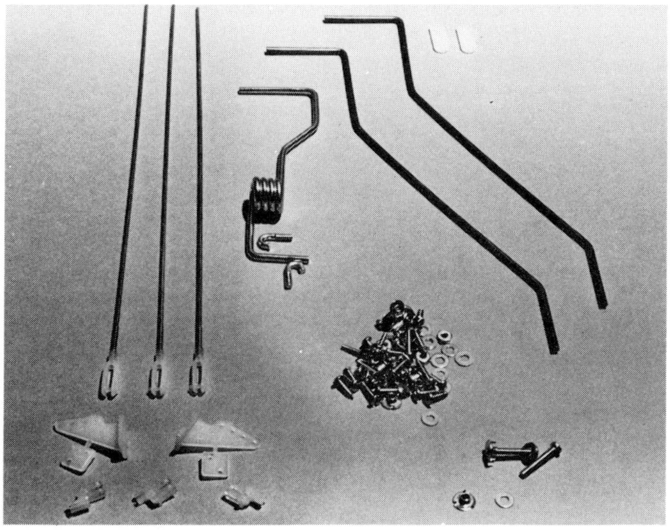

Some kits include installation accessories, so consider that when purchasing a kit. This hardware is part of the kit for the Goldberg Falcon 56 MK II, a popular trainer aircraft model.

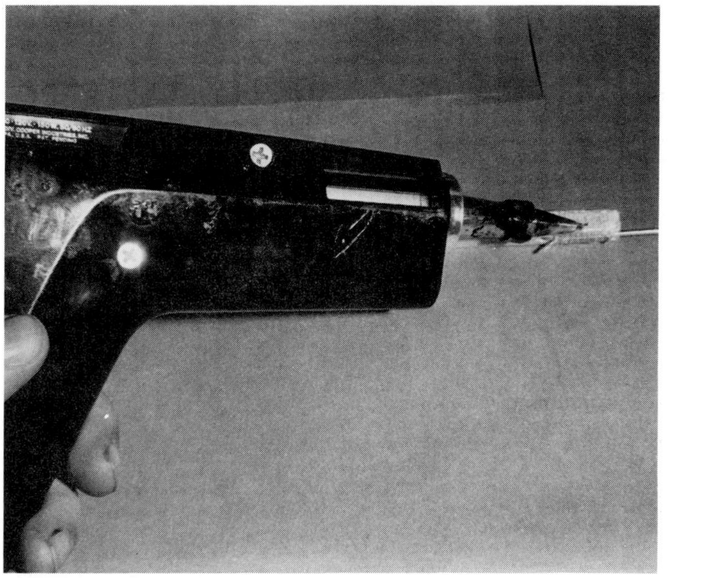

Thread and hot-melt glue speeds pushrod assembly.

Do this by reversing the two motor power leads and the two feedback pot end wires. Do not disturb the motor ground wires or the pot center tap wire. Please note that servo rotation will not be reversed by turning the servo around, although the direction of pushrod motion can be reversed by moving the pushrod clevis from one side of the output arm or disc to the other. The Ace Silver Seven, Kraft Signature Series, and Royal Omega transmitters permit servo reversal at the transmitter by the flip of a switch.

Examine the plans for the model you're building; some plans provide suggested installation information that may apply to your set. The equipment is usually placed to help balance the model. Do not, however, mount the equipment in some extreme position to help with the balance. For example, do not mount the receiver and servo against the fire wall. Balance weights are better and don't need repairs!

Determine the location of the receiver, servos, and batteries. In model aircraft, the battery is usually located in the fuel tank compartment. This means that the battery must be protected from fuel, first by wrapping the pack in ¼″ or ½″ thick foam rubber (not polyurethane foam), then by sealing the wrapped pack in a plastic bag. Close the end of the bag tightly with plastic tape or waxed cord. In an aircraft, the receiver and servos are usually located in the cabin for high-wing models, or in the area immediately above the wing for low-wing models. In boats and seaplanes the receiver and servos must be installed in a watertight box or built-in watertight compartment. Several manufacturers offer ready-made waterproof boxes or you can make your own from plexiglass and commercially available watertight fittings. In a car model, the entire set of equipment is usually mounted uncovered on the chassis. This is messy but there seems to be no alternative.

Having determined where the equipment is to be located, sketch on the plans the straightest runs for pushrods from control surfaces to servos. Determine the servo output motion required to produce the proper control surface or wheel motion. When working with low-wing aircraft, keep in mind that the plan view is different from the view seen when the fuselage is inverted; that is, the rudder and elevator movements must be carefully noted. Also, in an aircraft model, the nosewheel and rudder must turn in the proper directions. Figure 8-2 shows two methods of ensuring correct rudder and nosewheel movement. Holes through the fire wall and bulkheads must agree with the location of the engine throttle.

Use servo mounting trays whenever possible. Servos mounted in trays with grommets for vibration damping are easier to service and better protected than those mounted with double-stick servo tape. I've found that the tape provides little vibration damping, hardens with age, and sometimes comes unstuck.

Determine the type of pushrods to be used. Pushrods may be steel cable inside nylon sleeving, birch dowels, spruce or hard balsa sticks, or fiberglass arrow shafts. Occasionally, as on some ship models, you may want to use a double-cable system rather than pushrods. If in doubt as to what kind of pushrods are best suited to your model, consult the plans and use the material recommended by the kit designer. Determine the lengths of pushrods and the location of all exit holes, control horns, bell cranks, and other fittings. Sketch all these on the plans, preferably tracing the actual part to be used.

In planning, allow for adequate impact protection material around the receiver and battery pack. The foam rubber wrapping about these parts absorbs engine vibration very well, but does little to absorb impact loads. Therefore, install pads of Styrofoam or polyethylene foam in front of the airborne components. If possible, position the receiver so that its PC board lies flat against the impact-absorbing foam.

After you've completed planning, lay out the receiver, servos, pushrods, and other parts of the installation over the plans and check all clearances. Check again that the servos are oriented correctly.

Techniques

Parts needed for installations may include screws, nuts and bolts, hinges, control horns, pushrods, pushrod ends, pushrod keepers, clevises, brackets, bell cranks, throttle override devices, ball connectors, and steering arms. Several manufacturers, including Du-Bro and Goldberg, produce all these and most hobby shops stock them. Using the installation plan you have created, list all the items needed. Take the plan and the list to the hobby shop and ask the dealer to help you gather the necessary items. It's also wise to purchase an assortment of 2-56, 3-48, 4-40, and 6-32 nuts and bolts in several

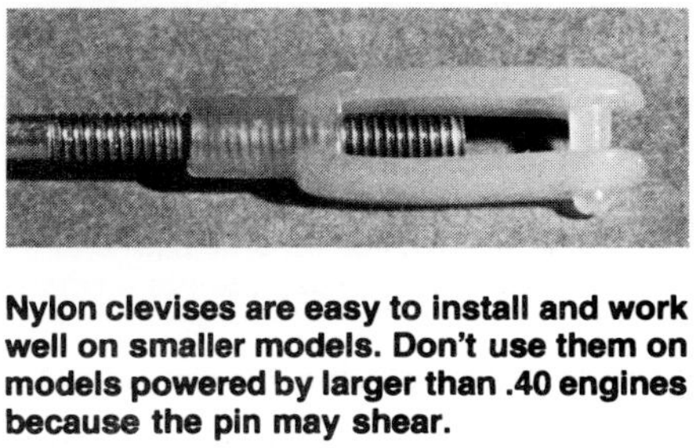

Nylon clevises are easy to install and work well on smaller models. Don't use them on models powered by larger than .40 engines because the pin may shear.

lengths. Sheet metal screws and wood screws may be necessary for some installations.

Install servos one at a time. Position the servo, assemble and install the pushrod, and connect all other parts. Then use the current measurement setup described in Chapter 6 to check that the servo does not draw more than 150 to 175 mA when it moves the control surface. For a rudder and nosewheel setup, make this test with the nosewheel held off the floor. Check current draw at both ends of full travel, including the increment added by trim. The most likely trouble spot is the throttle because it usually has solid stops at each end of travel. I've frequently heard a distinct growl from a servo at low throttle — a clear indication of incorrect installation. Use one of the manufactured throttle override devices or the homemade device, Fig. 8-3, to overcome this problem.

Other sources of high servo current draw include:

- Pushrods that rub against one another, have too much bend, or that fit too tightly at exit holes.
- Control surfaces that do not move freely or that jam at the end of travel.
- Pushrods and servo output discs that are misaligned with respect to one another.
- Servo output arms that rub against one another and pushrods that rub against the case of the servo.

Figure 8-4 shows examples of good and bad installation techniques that will help you avoid these and other problems. Lest you underestimate the importance of this, I once checked the current draw of the servos in an airplane model built by an experienced modeler and found that each servo drew 400 mA when exercised. For four servos, that totaled 1.6 A, enough to drain the battery in one short flight. He redid the whole installation after I drove him from my workshop at the rough end of a broom.

These checks will have verified that the installation isn't too tight. The next step is to be certain it isn't too sloppy. Loose linkages mean potential loss of control and can lead to flutter that will destroy servo gears and tear apart control surfaces. Hold each control surface and move the stick on the transmitter. The control surface should follow the control stick movement precisely with only the tiniest bit of play.

At the same time, check that pushrods do not buckle under the expected loads. These are conservative estimates of typical loads on the pushrod and servo:

- Small aircraft models up to .15-powered pylon racers: 2 to 3 pounds.
- Racing, pattern, and scale aircraft powered by up to .60 engines: 4 to 5 pounds.
- Quarter-scale aircraft powered by geared engines or converted chain saw engines: 6 to 10 pounds.
- Hydroplanes and other racing boats powered by .40 to .60 engines: 8 to 10 pounds.
- Other boats: 1 to 5 pounds.

This good installation could have been improved by reversing the direction of rotation of the lower left servo to avoid the kink in the throttle pushrod. Also, the nosewheel pushrod (upper left) should have shock-absorption. The fiberglass arrow shafts used as pushrods for the rudder and elevator are excellent.

Nylon sleeving must be secured with epoxy at both ends. Additional anchorages may also be required to prevent flexing. Flexible pushrods with nylon sleeving may cross over one another; other kinds may not.

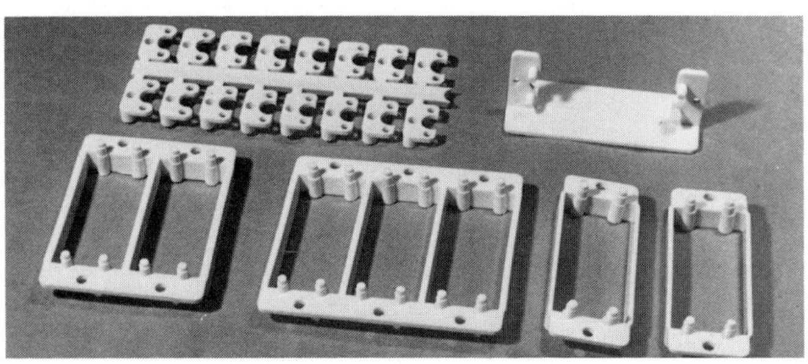

These servo trays accept one, two, and three servos and mount securely to wooden rails. Use trays whenever possible.

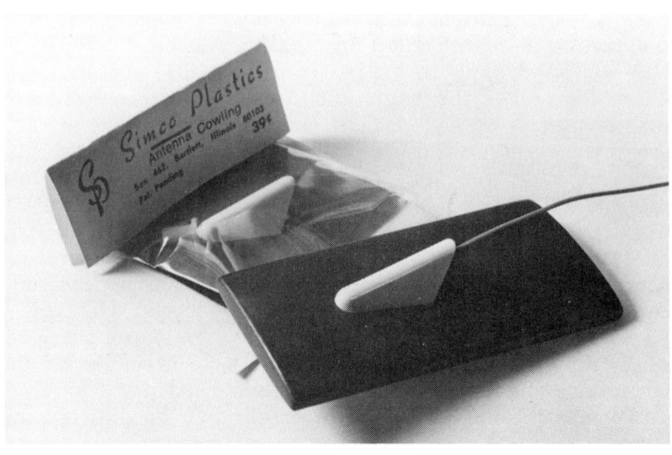

A simple device such as this can protect the antenna wire where it leaves the fuselage.

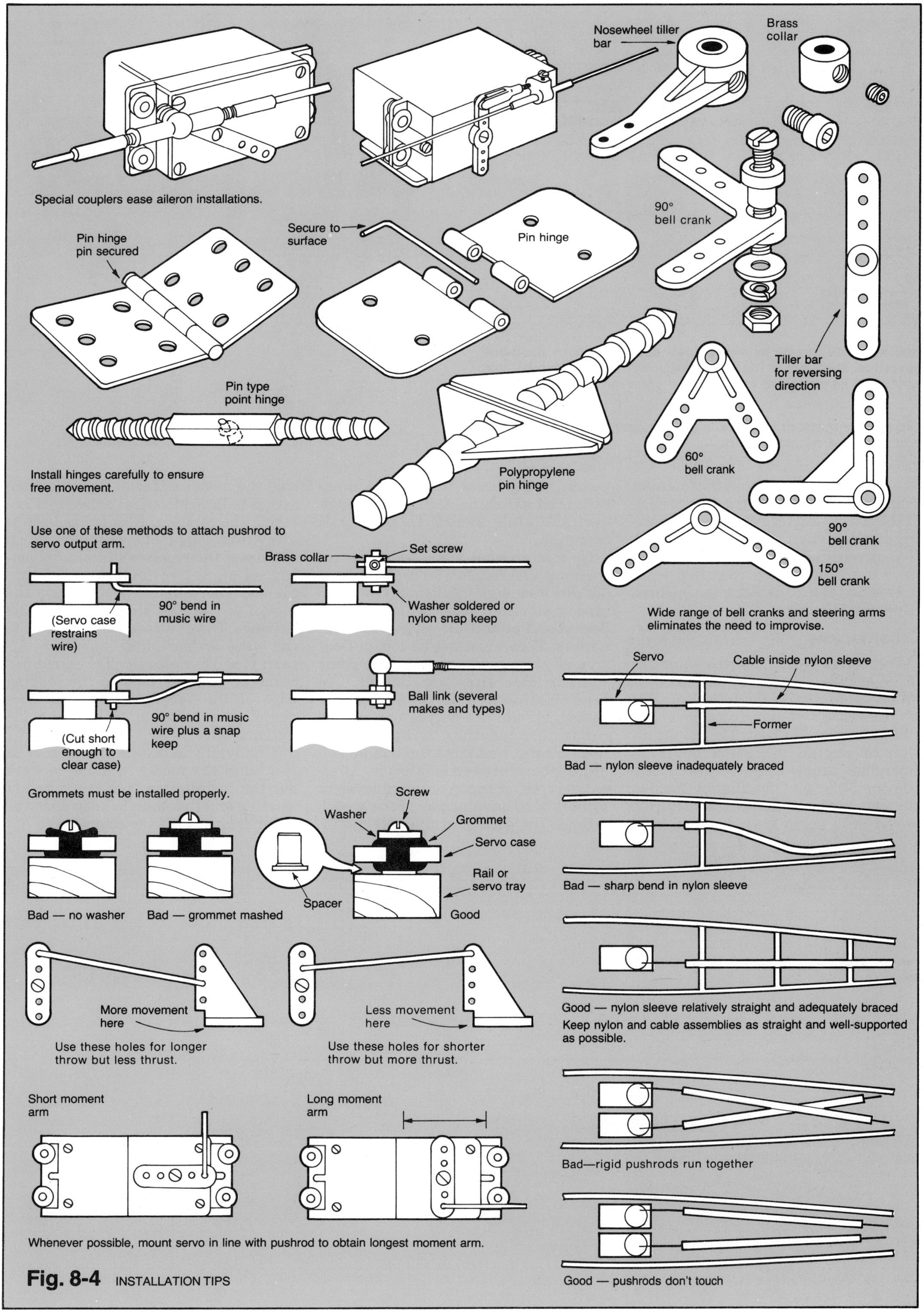

Fig. 8-4 INSTALLATION TIPS

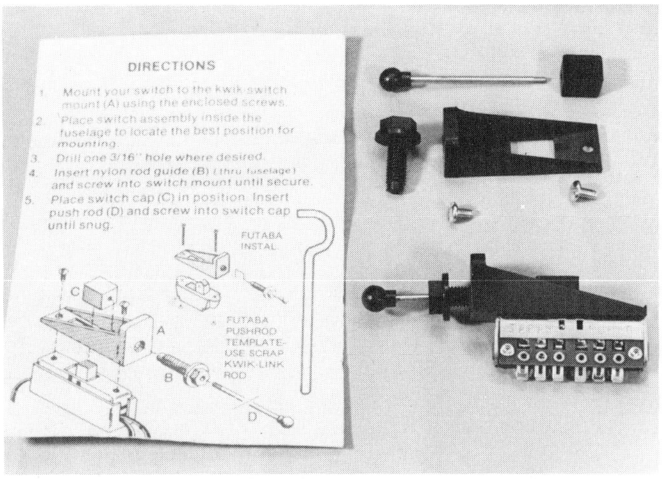

Use a switch mount to convert the slide switch to push-pull operation. Then mount the switch securely inside the model; that way it will be protected from dirt and vibration.

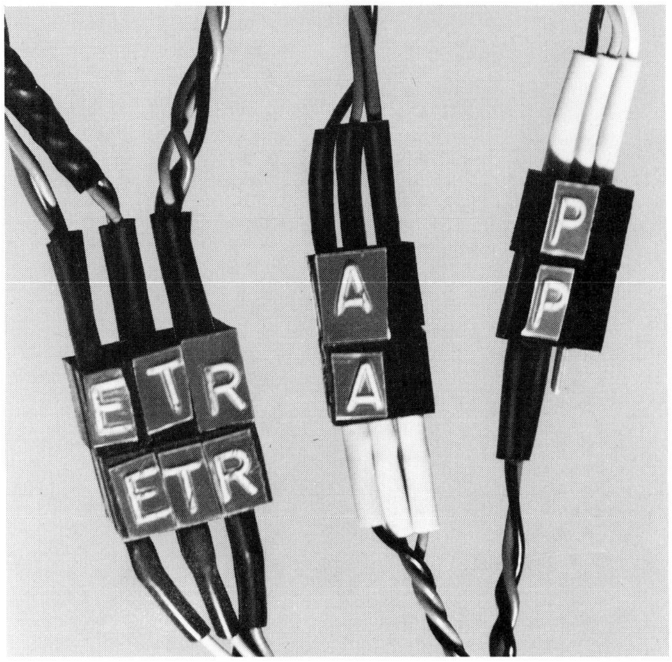

Label all connectors so that the landing gear won't retract when you move the aileron control.

These figures assume that the installation is correct, so that the free motion of the control surface and pushrod contributes less than a pound to the load. If your pushrods buckle when you apply a test load, change to stiffer pushrods or add fairleads along the pushrod to prevent flexing.

Other factors

The power switch is often subjected to unnecessary abuse by mounting it on the side of the model. Visualize a Quadra-powered quarter-scale fiberglass plane or boat. The amplification of engine vibration by large fiberglass panels could impose a continuous 50 to 60 G oscillatory load on the switch. Mount the switch inside the model on a servo tray or other solid mount, using brass tubing or a manufactured switch mounting kit to convert the slide switch to push-pull operation.

No matter what the model, seal all openings and hatches securely to keep out dirt and engine exhaust. Use foam tape or silicone rubber seals liberally. Secure and protect all wiring and connectors. They must not be left to float free so that connectors could come apart or loose wires jam controls. Use tape or clips to hold connectors together, bundle the excess wiring and connectors together, wrap them in foam plastic, and stuff the bundle solidly in place between bulkheads. Alternatively, tie or tape the wires to servo mounts or some solid part of the model.

Route the portion of the antenna inside the model as far as possible from servos to keep electrical noise from entering the receiver. Also try to keep it away from long metal pushrods which can upset the receiver's tuning by acting as signal reflectors or ground planes. The antenna should exit the model out of the side or top of the fuselage. Never attach the loose end to the wing or horizontal stabilizer. Attach it with a plastic clevis, rubber band, or piece of tape to the leading edge of the vertical stabilizer. Once again, never shorten or lengthen the antenna wire.

Finally, if you install the RC gear as you build the model in order to work out the installation, remove it before sanding and finishing to keep dust and paint splatters off the components.

9 Accessories

The development of accessories for RC modelers parallels the development of installation aids. In the old days we headed for the flying field with a cardboard box loaded with a No. 6 dry cell to heat the glow plug, a plastic squeeze bottle or rubber fueling bulb to fill the tank and prime the engine, a winder to wind up the rubber band that drove the escapement, extra props, a few tools, and the mandatory VOM for checks of battery voltages and receiver tuning.

Today, thanks to advances in technology, the experienced modeler is likely to show up at the field with a plywood or fiberglass box that is almost a piece of furniture. The field box may contain a 12-volt battery that powers an electric starter, an electric fuel pump, a glow-plug driver, and a fast-charger to recharge nicads used to power models with electric motors. Also included may be an expanded-scale voltmeter to be sure the batteries are okay, a radio monitor to check for interference, and even an audio or optical tachometer to check engine rpm.

The model may contain a digital power control (for models that are electrically powered), a servo reverser, a servo signal inverter that lets a negative-going-pulse servo work with a positive-going-pulse servo and vice versa, and a sonic locator that triggers whenever the signal disappears so that a lost model can be found more easily.

The transmitter may contain a built-in ESV or a low-voltage warning alarm, an electronic mixer, and a timer to warn of imminent fuel exhaustion.

Each modeler has his own list of essential accessories and tastes vary widely. What one modeler thinks essential, another may consider useless. As is the case with RC systems, the batteries that power the accessories cause the most trouble and require the most maintenance, so this chapter will focus primarily on batteries and battery chargers.

Batteries

Carbon-zinc dry cells, nickel-cadmium batteries of both open and sealed construction, and lead-acid batteries with liquid or gelled electrolytes are the most commonly used types of batteries in RC accessories.

The No. 6 carbon-zinc 1.5-V dry cell is 2⅝″ in diameter and 6⅝″ tall. It is used primarily as a glow-plug heater. It is not rechargeable, though its life can be extended by storing the cell in a refrigerator or other cool place when it's not in use. It also helps not to short the leads. Indeed, the assistant who carefully wrapped the leads around two of my brand new No. 6 cells, then clipped the leads together to hold them in place, will never be forgotten.

Automotive and garden tractor liquid-electrolyte lead-acid batteries are used to power electric motors on glider launching winches, and are occasionally used to power engine starters. Engine starters are more frequently powered by liquid-electrolyte lead-acid motorcycle batteries. Sealed lead-acid batteries with a gelled electrolyte can be used in any position and are safer than liquid-electrolyte cells because the battery acid cannot spill out. They are available in several sizes and can substitute for liquid-electrolyte lead-acid batteries in all applications.

In addition to their uses in RC transmitters and receiver/servo units, nickel-cadmium batteries are used to power glow-plug heaters and engine starter batteries. Many other accessories, such as electronic timers and low-voltage detectors, are powered by nicads. Electric-powered planes, cars, and boats often use nicads as their energy source.

Chargers

Both lead-acid and nicad batteries can be charged from another battery or from a charger powered by 110-volt household current. In either case, the battery is supplied a specified current for a specified time. For nicads, charging current is usually one-tenth of the battery's mAh rating and time is usually 14 to 16 hours. The hours beyond 10 are to make up for losses due to the battery's internal resistance and other factors.

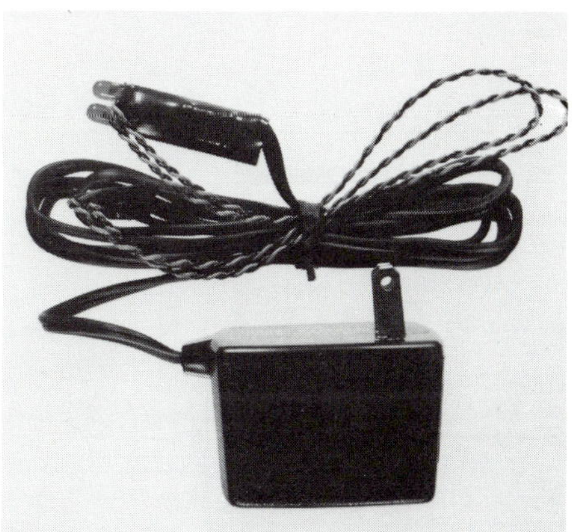

The simplest commercial charger contains a transformer, diode, dropping resistor, and an LED. This type of charger is usually supplied with RC systems.

Figure 9-1, which shows typical charging curves for nickel-cadmium batteries, and Fig. 9-2, which shows the curve for a lead-acid battery, reveal that charging rates are not linear. Further, both kinds of batteries are damaged by overcharging.

The three types of chargers in general use are those using a fixed charging rate, those using a variable charging rate, and those incorporating meters that allow the output to be monitored. Fixed-rate chargers are supplied along with the nicad battery packs in RC systems and are also available separately for other batteries. Their circuitry and parts layout, Fig. 9-3, is remarkably simple. The upper diode, the 150-ohm resistor, the LED, and the 4.8-V output from the upper portion of the circuit charge the receiver battery; in the lower circuit the 82-ohm resistor supplies 9.6 V to the transmitter battery. Similar circuits can handle battery charging tasks for packs ranging from 2.4 V to 12 V and

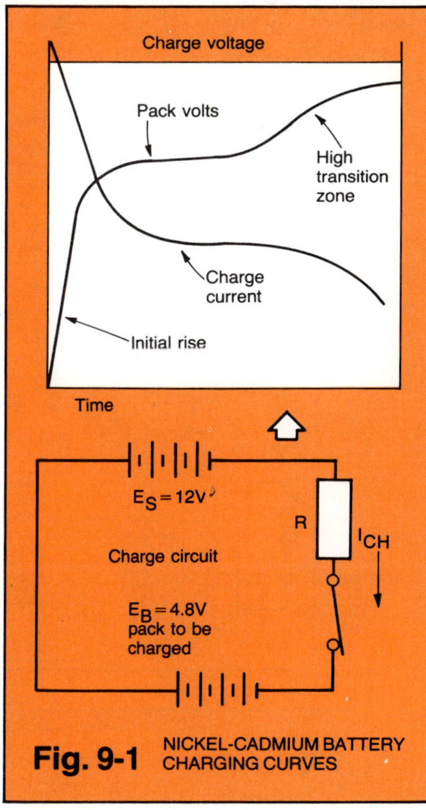

Fig. 9-1 NICKEL-CADMIUM BATTERY CHARGING CURVES

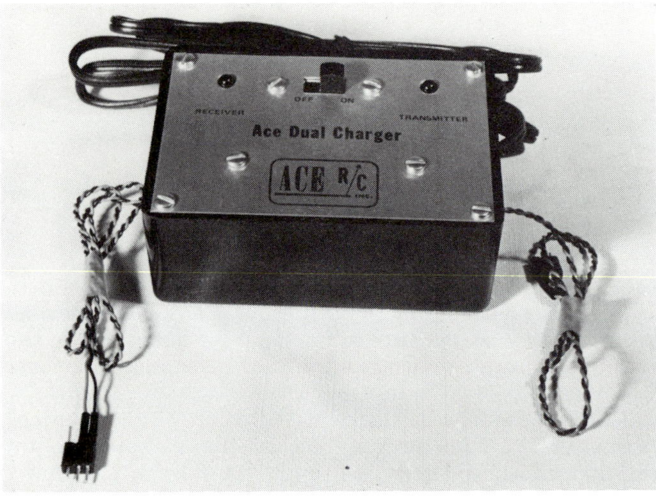

The Ace Dual Charger simultaneously charges 4.8-V and 9.6-V battery packs.

The inside of the Dual Charger reveals its simple design. Never operate the charger outside its case because 110 VAC is present at the switch and transformer inputs.

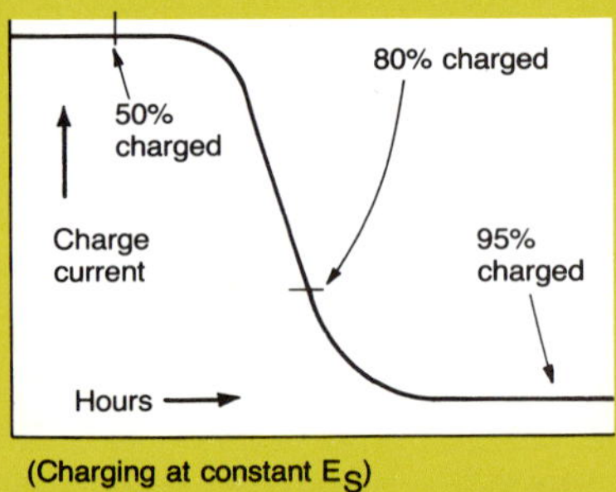

Fig. 9-2 LEAD-ACID BATTERY CHARGING CURVE

for charging rates from 10 mA to 50 mA. If a potentiometer replaces the fixed-value resistors of Fig. 9-3, the charger's output voltage can be varied to meet the needs of different batteries, Fig. 9-4. We now have a variable-rate charger. Adding an ammeter allows the output to be monitored so that cells may be charged at any desired rate.

More complex variable-rate metered chargers can handle nearly all battery charging assignments. Figure 9-5 shows the schematic for a charger I built from scrap box parts in 1963 and have used ever since. I won't show you a photo of it because it's too ugly. Beauty is only skin deep, but ugliness goes right to the bone! The charger is metered on any one of three outputs (I was too cheap to buy three meters) and has three charging ranges: 0 to 100 mA, 0 to 1 A, and 0 to 10 A. Because it was on hand, I used a transformer capable of handling 20 A. For most purposes, a 3- to 5-A, 18- to 28-V transformer will suffice. Diodes, resistors (or pots), and meters must be selected to handle the anticipated charging currents; obviously, if you put a 750-mA diode in a 10-A circuit, it will glow brightly for but a second. The values chosen for the meter circuit depend upon the meter purchased; the one I used happened to be inexpensive and has rather high resistance. Because the meter is part of the circuit, charging current increases when the meter is switched out. Therefore, calibrate the metering circuit. With an external ammeter in the circuit, check the charging current with the meter switched in and out. If the charging current increases more than 5 percent when the meter is switched out, a calibration factor must be noted on the charger and applied when setting charging current. For example, if you are charging a 450-mAh pack at 45 mA with the meter switched in and the current decreases to 35 mA when the meter is switched out, a 10-mA bias has to be applied when setting charging current. Note that the meter must be in the circuit for the 0 to 100 mA range, so it must always be switched back there after checking current on the two higher ranges.

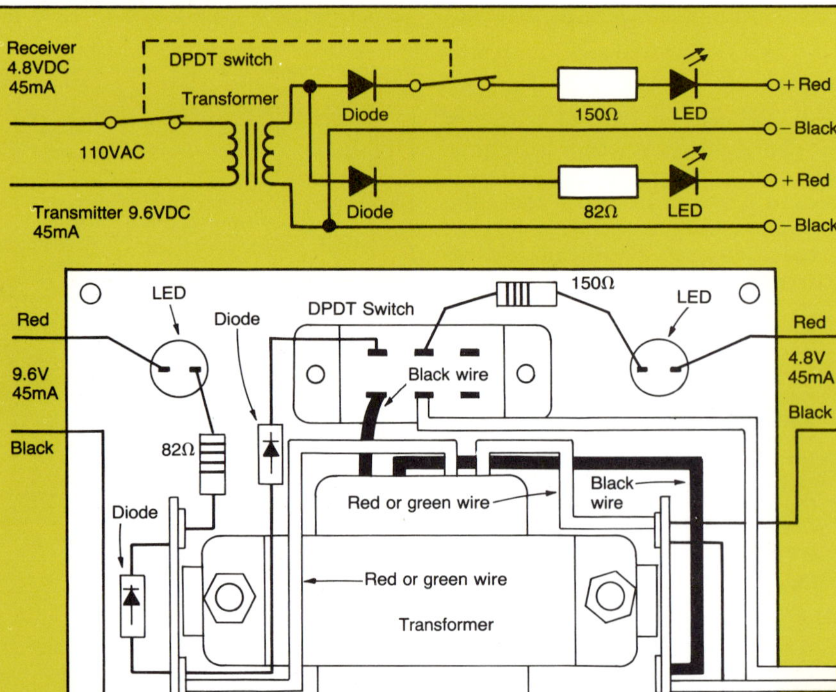

Fig. 9-3 DUAL-OUTPUT CHARGER

Fast chargers

General Electric and Sanyo have developed nicads with low internal resistance that can be charged in less than an hour. (Ordinary nicads have high internal resistance and can't dissipate the heat that would be generated by fast-charging.) These fast-charge cells are used in battery packs for electric

The Royal Electronics variable charger includes a potentiometer.

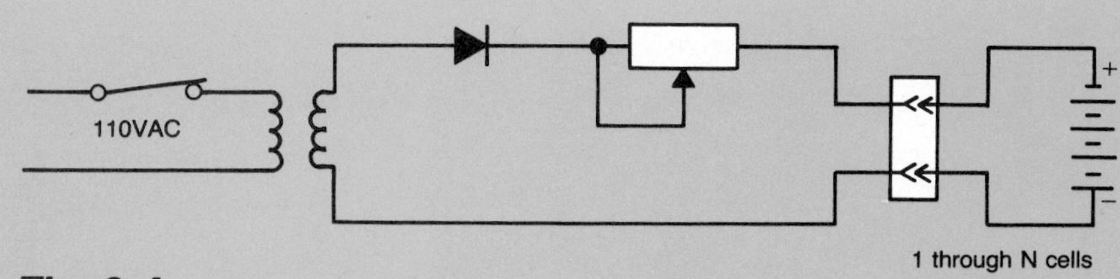

Fig. 9-4 PRINCIPLE OF THE VARIABLE-RATE CHARGER

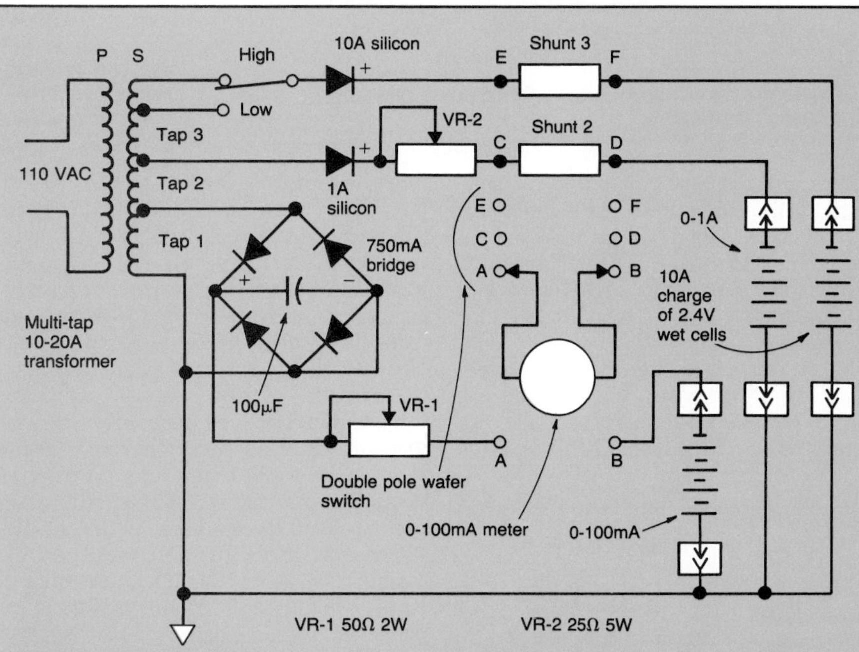

Fig. 9-5 TRIPLE-RANGE VARIABLE-RATE CHARGER

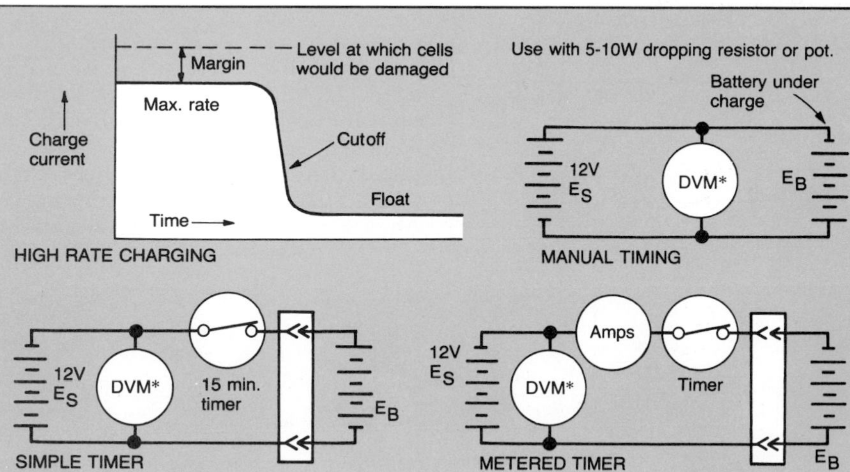

Fig. 9-6 TIMED RAMP-CHARGERS

*or ESV

cars and planes, and wherever else the combination of low weight, high storage capacity, and fast recharge capability is desirable.

Two types of chargers can be used with fast-charge nicads, the timed ramp-charger and the automatic-cutoff charger. Let's discuss timed ramp-chargers first. In the simplest version, a 12-V automobile battery or the 12-V output from a transformer-isolated DC power supply powers the charger, which also features a current-limiting resistor and a 15-minute mechanical timer. An ammeter or ESV should also be included, so that the charging process can be monitored. When using a timed ramp-charger note that the cells must be fully discharged to the 1.1 V per cell end point before recharging. This is nearly always the case for batteries used in electric cars that operate until the battery is depleted. It is not the case for receiver battery packs; discharge these before charging or you'll almost certainly overcharge and ruin them. Determine charging current by multiplying the battery's ampere-hour rating by 4 or 5. For example, the charging current for a 1.2-Ah cell should be 4.8 to 6 A.

Figure 9-6 presents schematics for three typical timed ramp-chargers and also shows typical fast-charge current curves. The proper value for the fixed resistor should be determined by using an ammeter in the circuit at the start of the charge. Alternatively, use a potentiometer and adjust according to the ammeter reading. The fixed resistor or pot should have a rating of 5 to 10 watts; determine the exact wattage after you've measured the fixed resistance and the current flow by using the formula $W = I^2R$. No resistance values are given here because they vary with the voltage rating of the battery, the charger supply voltage, and the maximum charging current. Be aware that the limiting resistor will get hot during charging, so mount it in a metal box to prevent burns.

In normal operation the initial charging current is 4 to 5 C, decreasing as battery voltage rises and the cell approaches full charge. If the battery is depleted when put on charge and the current exceeds 4 to 5 C, the pack probably has a shorted cell. If the charging current is zero, a cell is open. If it is not zero, but too low, a cell may have developed a high internal resistance. Replace any cells that have abnormal readings.

An ESV in the charger circuit or used in conjunction with the charger lets you monitor battery voltage. For example, a depleted 4.8-V pack might measure 4.4 V at the start of charge. As charging starts, battery voltage rises rapidly. After 3 to 5 minutes, voltage levels off, rising slowly to 5.5 V to 5.8 V (depending on the design of the

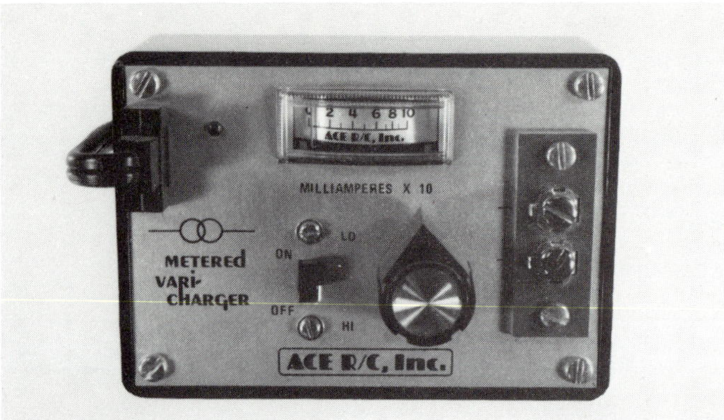

Ace's Metered Vari-Charger allows continuous monitoring of the entire charging process.

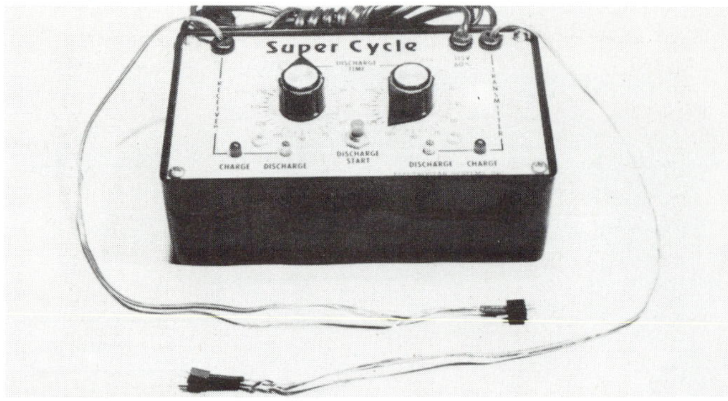

I've used an Electrostar Super Cycle for years. It simultaneously charges receiver and transmitter battery packs and uses two timers to record discharge time.

battery and the ambient temperature), where it should remain for some time. Any rise from the 5.5 V to 5.8 V range through 6 V toward 6.1 V indicates imminent overcharge and calls for immediate shutoff. Never let the voltage exceed 6.169 V.

The timer guards against such overcharging. For example, assume we want to charge a 6-cell, 1.2-Ah race car battery pack. The current limit is 4.5 A and the timer is set for 15 minutes. If current flowed at 4.5 A for .25 hour, the pack would receive 1.125 Ah of input, or 1.125/1.2 = almost 95%, dangerously close to an overcharge. However, as I mentioned earlier, current flow falls off as battery voltage rises, so that current is about 1 A after 15 minutes. Thus, the precise input isn't known because it's really the integral of the area beneath the current-time curve. Taking the liberty of straightening the curve, for our example 1.2-Ah battery a start current of 4 A and end current of 1 A yields a total input of 1 A times .25 hour plus .5 times 3 A times .25 hour, or .625 Ah, a conservative 50 percent of full capacity. That's okay for a short race, but I've known drivers to charge to 90 percent of capacity for long races. Be prepared to buy batteries frequently if you do this. Figure 9-7 further illustrates the charging curve, this time assuming we want to charge GE 550-mAh Sub C cells at an initial current of 4 C. If we again straighten the curve, with intercept at t = 0 of 1.8 A and with end point of .5 A after .5 hour plus the area of the triangle (.5 times .5 hour times 1.3 A), the total charge current is .575 Ah. These GE cells are conservatively rated, so can withstand this slight overcharge.

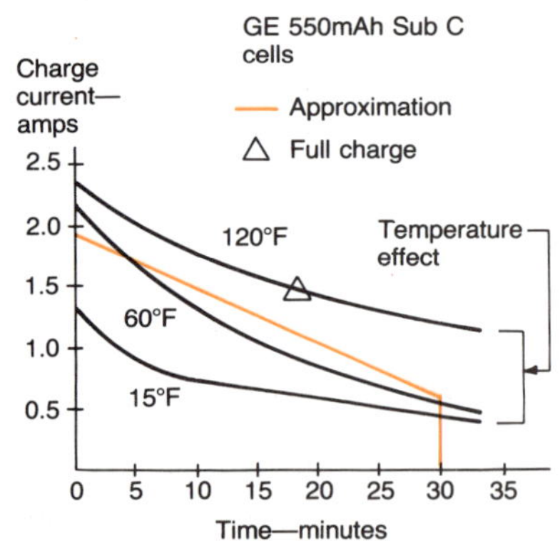

Fig. 9-7 FAST-CHARGE NICAD CHARGE CURVES

All of the chargers described so far rely upon you or a timer to take the batteries off the charger before they become overcharged. It would be nice to have a charger that delivers full charging current until just before the battery reaches full charge and then switches off or goes from full charging current to a float charge (just enough to keep the battery fully charged). Current solid-state technology makes such devices economically practical for hobbyists. Figure 9-8 illustrates a circuit from a National Semiconductor applications note for an automatic-cutoff charger suitable for use with gelled-electrolyte lead-acid batteries. As the battery charges and the battery voltage increases, more current flows through the regulator feedback circuit, effectively biasing the 2N5294 off and reducing current flow. The values in the circuit must be set up to accommodate the particular type of battery to be charged, but the general arrangement is the same for all types of batteries, including fast-charge nicads.

Not all automatic-cutoff chargers are compensated for changes in nicad charging characteristics at extreme temperatures. This means that an uncompensated automatic-cutoff fast-charger might not cut off on a very hot day, overcharging the batteries. Conversely, on a very cold day, it might cut off too soon, leaving the batteries only partially charged. Carefully monitor the batteries when charging nicads at temperatures lower than 40° F or higher than 90° F. Buy a timed fast-charger or automatic-cutoff fast-charger on the basis of specifications, not price. Study specs carefully, and purchase a unit that meets your needs.

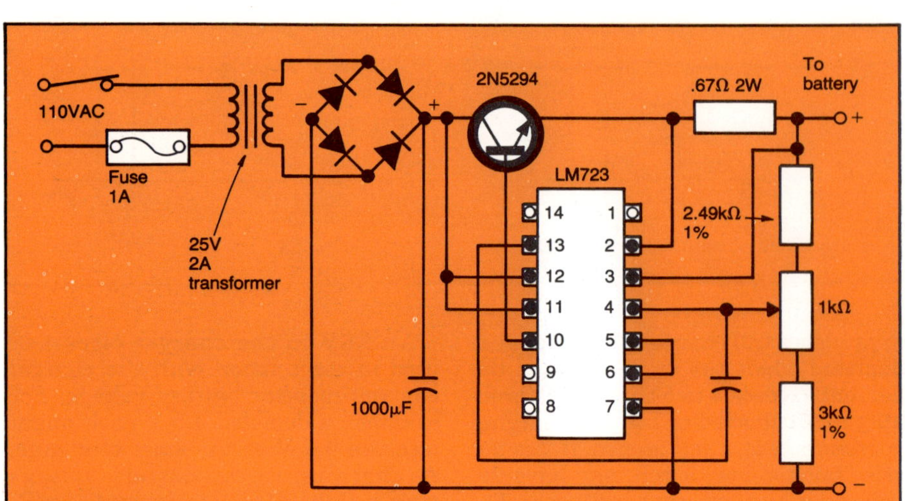

Fig. 9-8 AUTOMATIC-CUTOFF CHARGER

Expanded-scale voltmeters

I've mentioned ESVs so often that you've probably concluded they are among the first accessories you should purchase. You are correct. ESVs range from a simple unit from Ace, the Mini ESV, to complex units capable of han-

The Ace Digipace 1 is an extremely sophisticated charger and cycler featuring digital readout of discharge time for receiver and transmitter battery packs.

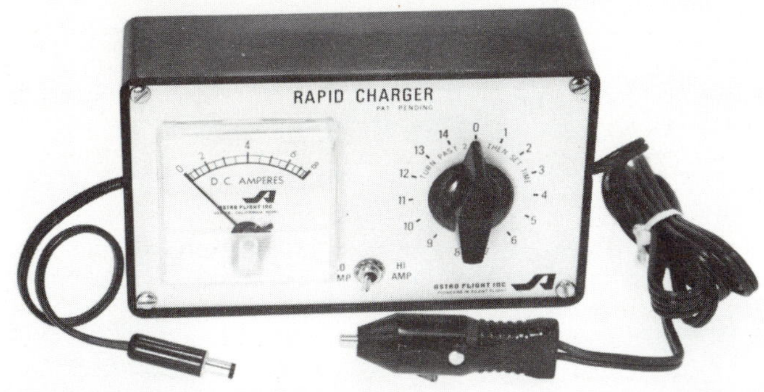

The Astro Flight Rapid Charger is a timed fast-charger for 1.2-Ah fast-charge nicads used in many model cars and aircraft. Powered by a 12-V automobile battery, it charges the 1.2-Ah packs to 80 percent of capacity in 15 minutes.

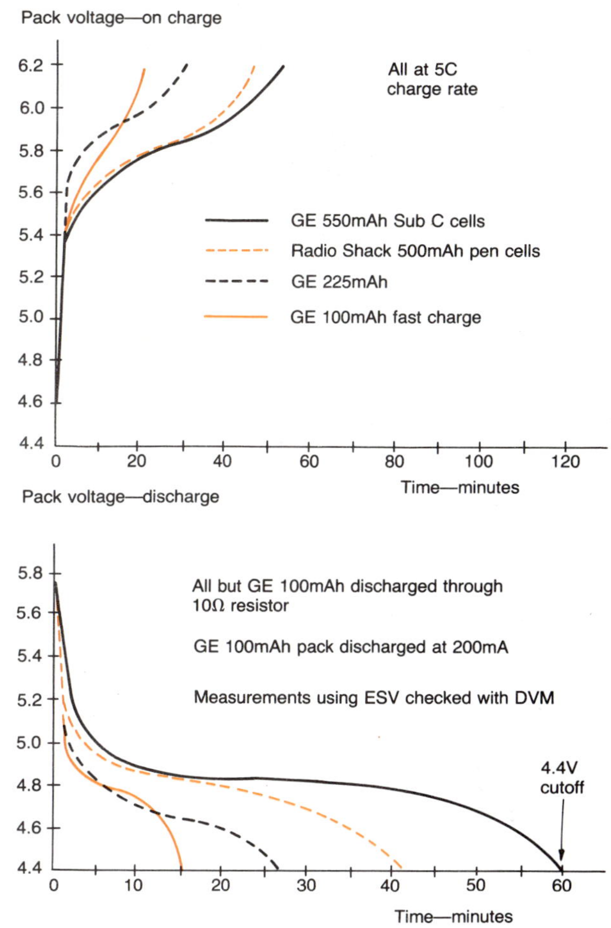

Fig. 9-9 NICKEL-CADMIUM BATTERY PERFORMANCE CURVES

dling the most complicated measurements. An ESV displays the voltage of a battery about its rated nominal voltage. For example, a 4.8-V ESV goes to full right scale deflection when the voltage exceeds 5 V and goes to full left scale deflection when the voltage decreases to 4.2 V. This is about a 5 to 1 improvement over the sensitivity of other types of voltmeters.

The ESV is most often used to measure battery voltage as a pack is discharged or charged. It can provide an accurate time history of charging and discharging voltages to a tenth of a volt. Assume that you've bought a new 4.8-V 100-mAh pack and you want to find out its capacity and discharge characteristics. Charge the battery to full capacity, then discharge through a measured resistance. Record the meter reading at 5-minute intervals and plot voltage as a function of time. The time required for the battery to discharge to 4.4 V is a measure of capacity. Use the curves in Fig. 9-9 for comparison. If the current drain through the fixed load is measured in milliamperes and plotted along with voltage, the area under the curve may be determined by counting

This Northeast Engineering Plugdriver uses 12-V pulsed DC to heat glow plugs to desired temperature.

The interior of the Northeast Plugdriver shows that it is controlled by an IC. A large heat-sinked transistor applies the voltage pulses to the glow plug.

grids to add up the total milliampere-hours between full voltage and 4.4 V. Use the same process for any other nicad.

The Ace Mini ESV is typical of simple ESVs. Figure 9-10 shows the schematic for the 9-V model. In this model D1 is a 6.8-V zener diode. The zener diode passes current only when the voltage reaches 6.8 V, and it conducts essentially full current flow at 7 V. R2 shunts most of that current. However, R2 is paralleled by R1 and the meter which has a nominal 1 mA flow for full scale deflection. In this example, the meter deflects full scale at 9 V and ESV current flow of 18 mA, 1 mA of which goes through the meter. At 7.3 V, ESV current drops to 2 mA and the meter is at zero scale.

More sophisticated ESVs use a differential method in which the test voltage is compared to a fixed reference voltage produced by mercury or alkaline cells inside the ESV. Figure 9-11, a design by Jim Oddino, shows the schematic for a differential ESV designed for 4.8 V. The meter is an Asahi HN 18 available from Ace. The reference cells provide a "bucking" voltage of 4.5 V. The input voltage is that of the battery being tested. The potentiometer is used to set the total resistance of the meter and pot so that the meter swings full scale at a battery voltage of 5.5 V, that is, when the differential between the reference cells and the test battery is 1 V and the current flow is 1 mA. When the test battery has dropped to 4.5 V, the differential is zero and the meter is at zero deflection.

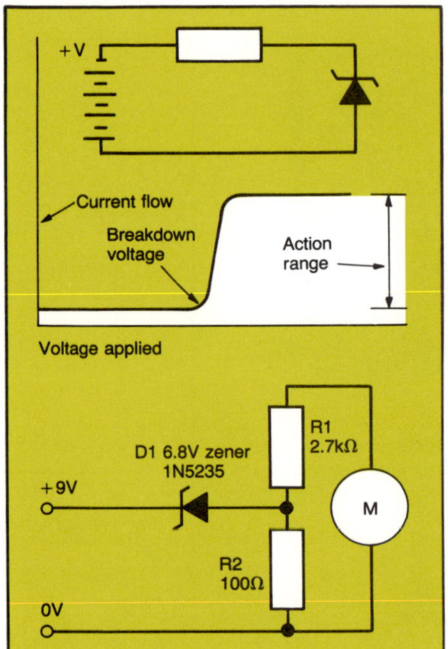

Fig. 9-10 ACE MINI-ESV

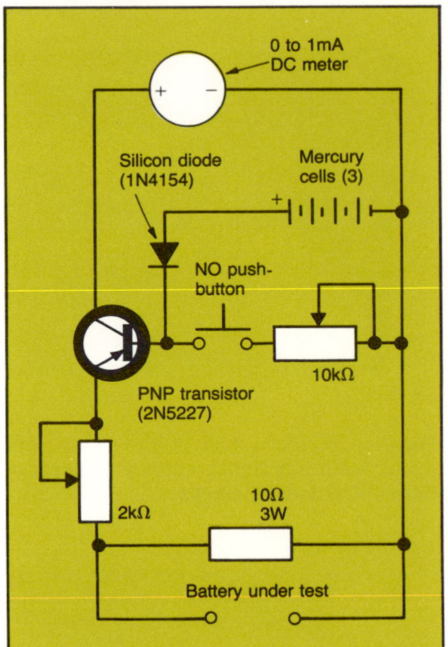

Fig. 9-11 DIFFERENTIAL ESV

This Ace servo reverser uses an NE544 IC and connects with the servo input wiring.

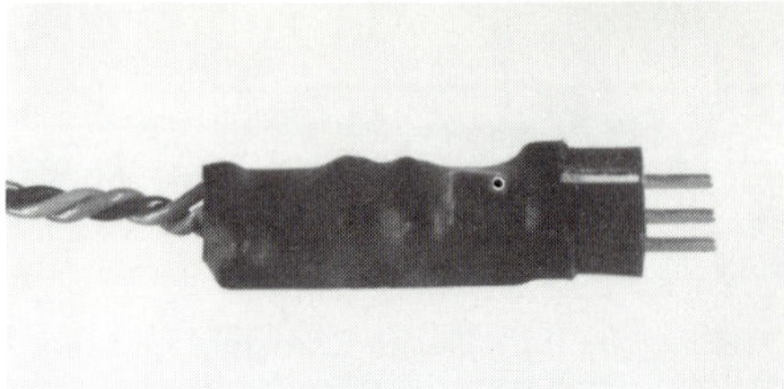

A servo pulse inverter attaches to the servo connector plug so that positive-pulse servos can be used with negative-pulse receiver/decoders and vice versa.

Other accessories

As I've mentioned earlier, a battery cycler simplifies nicad maintenance by automating the process, freeing you from the need to monitor with clocks and meters. Servo reversers and servo signal inverters are sometimes handy. A servo reverser allows you to change the output motion of the servo to suit special installation needs. Servo signal inverters let you use positive-pulse servos with negative-pulse receiver/decoders, and vice versa. In addition, voltage converter units are popular for field box use. These allow you to use a single 12-V battery to power a range of accessories and chargers, each with a different voltage and current requirement. Such a converter may include outlets of 2 V for glow-plug heating, 3 V for a fuel pump, and 12 V for a starter.

If you feel the need for some other accessory, such as a sonic locator that sounds an alarm whenever the receiver loses the transmitted RF and that serves to help you locate a downed airplane, search the ads in the modeling magazines: Chances are that several manufacturers offer the item.

This low-voltage warning device attaches to your transmitter and warns of low transmitter battery voltage.

Symbols

American National Standard electronics symbols are used in the schematics and block diagrams. The most common symbols are shown here.

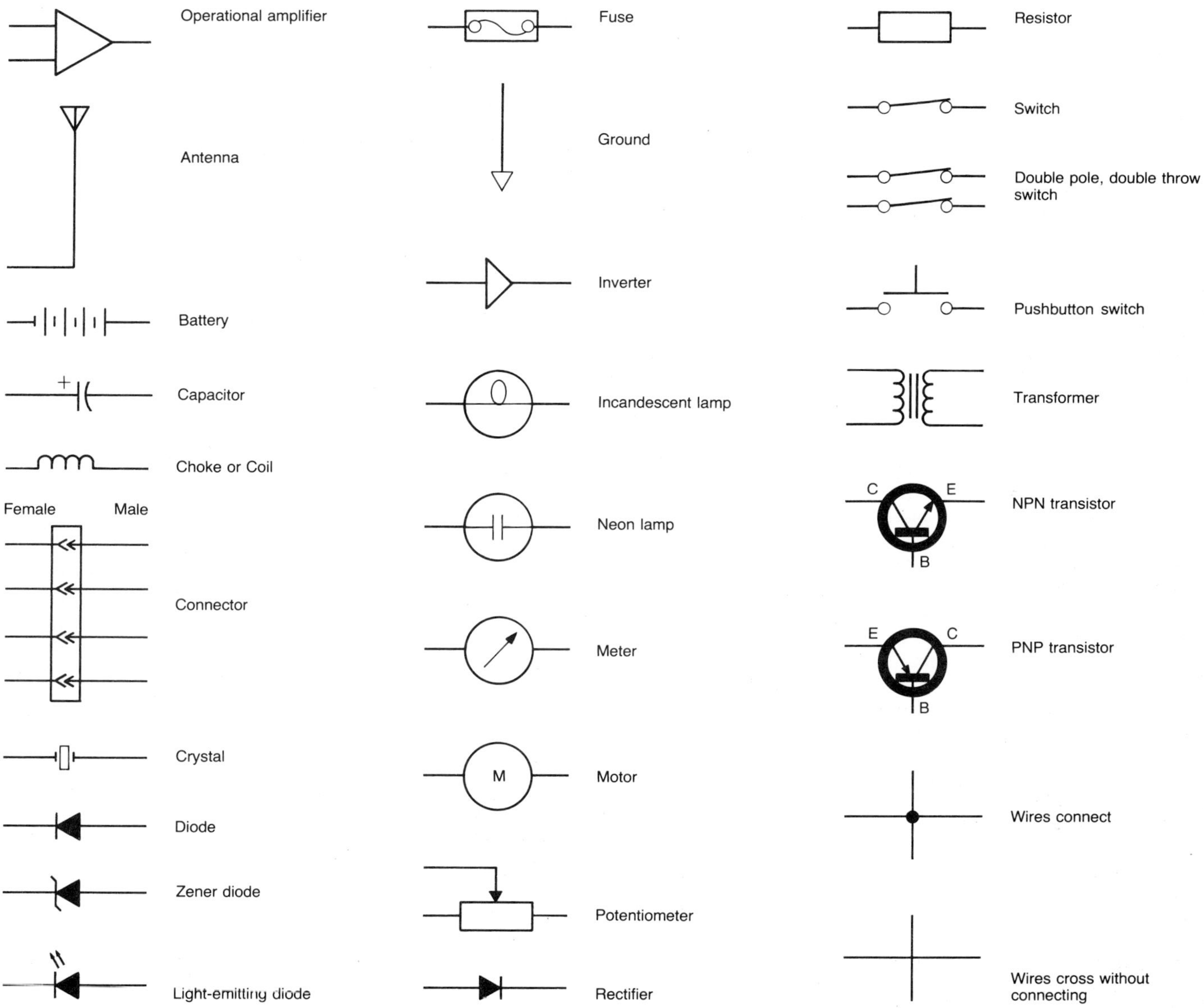

Acknowledgments

First, special acknowledgment goes to my family for their long patience and assistance in the preparation of this book. Daughter Diana did extensive literature research and she and my wife Patsy typed all the draft material. Sons James and Timothy helped with numerous projects and the photo sessions.

Many other persons provided information, advice, and materials. Special thanks go to Frank Pierce and Bill McCaw for guidance on photo techniques and many hours of darkroom work; to Paul Runge of Ace R/C for materials and photos of Ace systems; to Bill Cannon for photos and permission to quote his discussion of an early Cannon decoder; to Tom Gist, George Myers, and Jim Oddino, whom I've quoted or paraphrased in this book; and to Kraft Systems, Royal Electronics, and Jim Martin of Hobby Lobby for photos and schematics.

Finally, the experience and information generated over the years by the pioneers of RC are acknowledged. The information presented here would not have been possible without them. I am happy to have counted many of these fine people among my friends and acquaintances. Included are Paul Runge, Phil Kraft, Doug Spreng, Bill Cannon, Al Doig, Ed Lorentz, Bill Winter, Walt Good, Don Mathes, Dick Rehling, Bob and Dick Dunham, Jack Port, Don Dickerson, Harry Sampey, Bob Elliot, Sid Gates, Bill Hershberger, and Howard McEntee.

About the author

Fred M. Marks built his first model plane at the age of 6, and by the age of 12 he was convinced that he would become an aeronautical engineer. Fred received his degree in aeronautical engineering from West Virginia University in 1957. He has worked for Boeing, Fairchild, and ITT, and has been a consultant with Booz, Allen and Hamilton, Inc. for 15 years, where his work often involves helping such clients as the U.S. Navy solve problems with full-size aircraft and their support systems.

Fred entered the RC hobby in 1959 after having flown free flight and control-line models for many years. By 1961, he was designing and fabricating his own equipment. By 1968, Fred had written the first of numerous articles for modeling magazines. During the early 1970's, he designed several RC systems, including digital feedback sets. He has coauthored *RC Systems for Model Builders* and BASICS OF RADIO CONTROL MODELING.

Fred has watched the steady growth of the technology of RC systems and the tremendous influx of newcomers to the hobby. He was prompted to write GETTING THE MOST FROM RADIO CONTROL SYSTEMS by a desire to help beginners as well as experienced modelers enjoy the RC hobby safely and economically.

Fred lives in Gaithersburg, Maryland, with his wife Patsy and children Diana, James, and Timothy. Fred is a Deacon in the First Baptist Church of Gaithersburg, so all Sunday flying is in the afternoon.

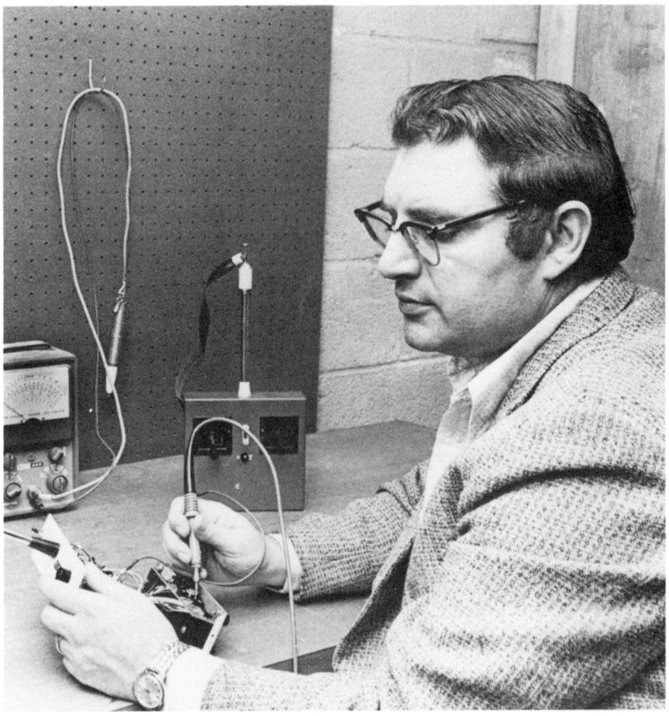